SCIENCE
ON TRIAL

SCIENCE ON TRIAL

THE CASE FOR EVOLUTION

DOUGLAS J. FUTUYMA

PANTHEON BOOKS, NEW YORK

Library of Congress Cataloging in Publication Data
Futuyma, Douglas J., 1942–
    Science on trial.
    1. Evolution. I. Title.
QH366.2.F88 1983   575   82–47880
ISBN 0–394–52371–7   AACR2
ISBN 0–394–70679–X (pbk.)

Manufactured in the United States of America

First Edition

Text Design: Robert Bull

To the memory of Marston Bates and Nancy Bell Bates,
who showed so many the way to reason and freedom.

CONTENTS

SPECIAL NOTE
ON FOOTNOTES

Because the subject of this book is controversial, I have provided copious endnotes for each chapter, to serve as documentation. Where copyrighted publications of the Institute for Creation Research are quoted, I have adhered to the conditions stipulated by the publishers of this material: that the entire paragraph in which quoted material appears be reproduced. Where the flow of the text would be interrupted by quoting a paragraph in full, the balance of the paragraph is provided in the referenced endnote, so that any individual who wishes to see the full context of the quote may find it at the end of the book.

—Douglas J. Futuyma

PREFACE

Evolution has, by now, the status of fact. It is one of the most important discoveries of science, and one of the most profound concepts in Western thought; so it is a sad irony that, a century after Darwin's death, the creationist movement is stronger than ever, carried forward by the New Right's rise to power. Part of the reason for this is that many people simply do not know any of the evidence for evolution, and still think of evolution as "just a theory," because scientists have not been very active in disseminating their ideas and findings outside the scientific community.

In the following pages, I shall describe the evidence for evolution, explain how the evolutionary process is thought to operate, expose the fallacies in the arguments that creationists use, and place the controversy in a larger scientific and social context. I hope to show that the attack on evolution is an attack on science in general; that to accept the doctrine of creation rather than the evidence for evolution is to be guided by wishful thinking rather than by reason and sober judgment; and that the drive to establish creationism in the public schools is only part of a broader reactionary movement to supplant American pluralism with unquestioning adherence to the claims of authority and religious orthodoxy.

I have attempted to write this book for the reader who has little background in biology or in science generally. I hope it will prove interesting and useful not only for the general reader but for students and teachers as well. It is certainly not meant to be a textbook, but I have provided notes for documentation, and references for possible further reading. I do not expect to convert fundamentalist creationists to belief in evolution. Fortified against logic and evidence by unquestioned doctrine, they are not likely to be swayed. This book is addressed, rather, to the reader who is open to evidence on matters of scientific substance.

I am grateful to the graduate students and faculty members at Stony Brook, especially Michael Bell, Stefan Cover, Scott Ferson, and Jerry Hilbish, for sharing their ideas with me, and to Charles Mitter for spurring me to more rigorous thought. Bruce Smith and Judith Koehn were kind enough to read much of an early draft of the manuscript, and Norman Creel and William Jungers read chapter 5. I thank Joyce Schirmer for preparing illustrations, and Steve Haber, Rita Sickles, Gwen Bellino, and Wanda Mocarski for typing. Stephen Jay Gould and Philip Pochoda suggested that I embark on this project, and Tom Engelhardt's editorial abilities were indispensable. Mary McCallum's friendly help in amassing creationist literature is greatly appreciated. Finally, I am immensely grateful to Bruce G. Smith for unflagging advice and support.

SCIENCE
ON TRIAL

I have said that the man of science is the sworn interpreter of nature in the high court of reason. But of what avail is his honest speech, if ignorance is the assessor of the judge, and prejudice the foreman of the jury? I hardly know of a great physical truth whose universal reception has not been preceded by an epoch in which the most estimable persons have maintained that the phenomena investigated were directly dependent on the Divine Will, and that the attempt to investigate them was not only futile but blasphemous. And there is a wonderful tenacity of life about this sort of opposition to physical science. Crushed and maimed in every battle, it yet seems never to be slain; and after a hundred defeats it is at this day as rampant, though happily not so mischievous, as in the time of Galileo.

—THOMAS HENRY HUXLEY, 1860

ONE

REASON UNDER FIRE

Whatever additional factors may be added to natural selection—and Darwin himself admitted that there might be others—the theory of an evolution process in the formation of the universe and of animated nature is established, and the old theory of direct creation is gone forever. In place of it science has given us conceptions far more noble, and opened the way to an argument from design infinitely more beautiful than any ever developed by theology.
—ANDREW DICKSON WHITE, 1896*

In the fourth century A.D., St. Augustine declared that "nothing is to be accepted save on the authority of Scripture, since that authority is greater than all the powers of the human mind." "Moses opened his mouth," affirmed St. Ambrose, "and poured forth what God had said to him." And so the fathers of the church established one of the most

---

* Andrew Dickson White, the author of *A History of the Warfare of Science with Theology in Christendom*, was the first president of Cornell University.

powerful, lasting beliefs in Christian civilization: the literal truth of the Bible's every word.

Sixteen centuries later, Biblical scholarship and archaeological research have established that the story of creation with which the Bible opens was developed by the Hebrews from even more ancient Babylonian and Chaldean myths, and that it was written by at least four authors. In addition, in the last four centuries the literal interpretation of the Bible has yielded before the discoveries of science. We no longer believe, as the ecclesiastical authorities of the Middle Ages held, that the heavens are a solid crystalline vault from which the sun, moon, and stars are suspended; that rain falls through holes in that heavenly vault from a celestial cistern; that the sun and the heavenly bodies rotate about the earth; or that comets and storms are warnings from God or machinations of the devil. These beliefs were rooted in Scripture and held sway for centuries; and every one of them was defended bitterly by orthodox theology against the observations of science. Physics, mathematics, astronomy, meteorology, geology—each of these, in asserting the power of reason and observation to understand natural phenomena, was condemned as a materialistic, atheistic attack on Christian faith.

It is difficult to believe that in the last decades of the twentieth century, when we have sent spacecraft past Saturn, discovered the forces that move continents and lift mountains, traced the biochemical pathways of the cell, and revealed the molecular structure of the gene, science should still be at war with the remnants of medieval theology. But religious orthodoxy, while it has retreated before physics and chemistry, has still not come to terms with biology.

Darwin's theory of evolution was as threatening to orthodox theology when it was published in 1859 as Galileo's astronomy had been in 1615. It was the symbol of atheistic materialism in 1925, when the high school teacher John Scopes was found guilty of violating the Tennessee law against teaching evolution. Fifty-seven years after the Scopes trial, fundamentalist religion and evolutionary biology are again fiercely at odds, and science is still on trial.

It is under more serious attack now than it has been for half a century.[1] The threat is not trivial. By November 1981, two states had passed laws requiring creation to be given equal time with evolutionary science in public school science classes, and similar bills were under consideration in more than twenty other states. Creationist groups have drafted bills that they are circulating widely among the other state legislatures. Similar initiatives are under way in countless local school districts where boards of education are yielding, or resisting only with difficulty. Museums have come under attack for presenting displays on evolution. Some members of Congress have considered introducing a bill that would mandate federal support for creationist "research." And the fundamentalist assault is not limited to evolutionary biology: physics, astronomy, geology, anthropology, and psychology are all under attack. Nor is science the only intended victim. Creationists represent only one facet of a movement that is dedicated to extinguishing "secular humanism," under which they include all attitudes and educational programs that do not explicitly include their theological doctrines. As a leading creationist, Nell Segraves, has put it, "We have a lot to undo. Creation/evolution is only the beginning."[2]

The challenge to evolution that has been mounted by religious fundamentalists touches us all. Most immediately, of course, it is an attack on biologists, for evolution is the single most pervasive theme in biology, the unifying theme of the entire science. But physicists, too, will find themselves under fire: they may be able to discover the structure of the atom, but according to the fundamentalists, physicists are wrong in claiming that radioactive atoms break down at a constant rate. All of geology is under siege: the entire petroleum industry may be built on geological knowledge, but geologists' evidence of the earth's age and the forces that have shaped it is, according to the fundamentalists, all worthless. Astronomers may be able to measure the speed of stars billions of light years away, but when it comes to their evidence of the age and origin of the universe, they are all wrong. Anthropologists, too, are under fire: they not only teach evolution, but they commit the unforgivable sin of cultural relativism: describing the peculiar habits of different peoples without condemning their immorality. Linguistics is also anathema: the notion that human languages have developed from one another is an evolutionary doctrine that contradicts the Biblical story of the tower of Babel. In short, all the sciences are under attack. But it is not only the content of the sciences that is contested. The creationists are assaulting the entire mode of scientific thought and the guiding principle of science: that traditional beliefs are open to skeptical inquiry.

Creationism, of course, is not new. Until the middle of the nineteenth century, most Western Europeans, reared in the Judeo-Christian tradition, believed that species, including the human species, had each been created in their present form by an omnipotent God. This belief was as prevalent among scientists as among non-scientists. While there were a few who doubted the doctrine of special creation, the world of science did not fully embrace the idea of evolution until after 1859, when Charles Darwin published *The Origin of Species*. The belief in special creation, however, persisted among many non-scientists, who were unaware of the evidence for evolution, and whose beliefs were guided by religion. In Europe, religion has long had a fairly intellectual tradition, so the popular outburst against evolution subsided within a few decades after *The Origin of Species*, and evolution has not been a noticeable subject of public debate since then. But in the United States, fundamentalist religion, holding a literal interpretation of the Bible, has proved a more tenacious and powerful opponent, so there have been periodic outcries against teaching evolution in the public schools.

The trial of John Scopes in 1925 was one such confrontation between science and fundamentalist religion. Contrary to common belief, the Scopes trial was not a victory for evolutionary science. Scopes lost the case, because he had in fact violated the Tennessee law against teaching evolution (a law that was not repealed until 1967). In the wake of the trial, textbook publishers, afraid of losing sales, quietly reduced or eliminated any coverage of evolution in high school textbooks, and by 1942, less than half the high school science teachers in the country were teaching anything about evolution.[3]

When the Soviets launched Sputnik in 1957, the widespread fear that the Russians had outrun the United States in science and technology prompted a national reform of science curricula. The National Science Foundation began to sponsor the development of textbooks in biology that were written by professional biologists and emphasized evolution as the "warp and woof of modern biology." These immediately came under fire by fundamentalist groups, which charged, as in a 1973 case in Colorado, that the National Science Foundation books used federal funds to establish "secular humanism" as the official religion of the country. The suit was dismissed by the United States Supreme Court in 1975. In 1973, a new Tennessee statute required that textbooks treat evolution as a theory rather than a scientific fact, and that other theories, including Genesis, must be included in the biology curriculum. The National Association of Biology Teachers succeeded in getting that law repealed, on the grounds that it tended to establish religion in the schools.

Our present climate has become increasingly hospitable to such fundamentalist attacks. However, the attack on evolution is only one item in fundamentalism's program of battle against all social and intellectual change. With its rejection of Biblical scholarship and metaphorical interpretation of Scripture, fundamentalism appeals to emotion rather than reason, and its defense of "traditional" values and mores feeds on anti-intellectualism, conservatism, and fear of social change.

Thus the reaction against women's rights, gay rights, abortion, welfare, and pacifism that developed in the late 1970s and the 1980s has been associated with an increase in the strength and stridency of the fundamentalist religious right. Fundamentalism has also fed on the anti-rationalism that grew in the 1970s, when astrology, occultism, and religious cults flourished as they hadn't for decades. The political climate under the aegis of the New Right has found representation in a fundamentalist Secretary of the Interior who believes that Christ charged us to occupy the land until he returns and a President who has said that "if evolution is taught in the public schools, then the Biblical story of creation should also be taught." The pressure to include creation in science curricula is stronger now than it has been for fifty years. The fundamentalists, moreover, have learned from their past errors and are using more sophisticated tactics.

In the past, the vocal opponents of evolution were straightforward. Groups like the Jehovah's Witnesses argued openly that evolution was contradictory to their belief in the literal truth of the Bible. Any attempt to outlaw evolution from the schools or to teach the Biblical story of creation was therefore doomed to failure in the courts: it was a clear violation of the First Amendment of the Constitution, which the Supreme Court has interpreted to mean that neither a state nor the federal government "can pass laws which aid one religion, aid all religions, or prefer one religion over another" (*Everson* v. *Board of Education,* 1947). In the last ten years or so, however, some of the more wily creationists have adopted protective coloration. They call themselves "scientific creationists" and use the language of science to argue their case, while attempting to avoid any explicit reference to their religious beliefs and motives.

Who are these creationists? At present, one of the most active groups is the Creation Research Society in Michigan, whose voting members must have at least a master's degree in some field of science, but who must subscribe to the belief that the Bible is the written word of God and that all its assertions are historically and scientifically true.[4] In 1970 a splinter group of this society was formed in San Diego under the name of the Creation Science Research Center by Mrs. Nell Segraves. The stated purpose of this organization was "to take advantage of the tremendous opportunity that God has given us . . . to reach the 63 million children in the United States with the scientific teaching of Biblical creationism."[5] Mrs. Segraves is reported[6] to have said that she opposes the separation of church and state, that "there should be one nation under God," that children are sinful by nature, and that "we want 50 percent of the tax dollar used for education to our point of view."[7] She has briefed Ronald Reagan for years on the subject of evolution.[8]

Mrs. Segraves's son Kelly is the coauthor of *The Creation Explanation*,[9] which teaches that "the Christian student of origins approaches the evidence from geology and paleontology with the Biblical record in mind, interpreting the evidence in accord with the facts of the Bible." Mr. Segraves, with the help of Richard Turner, a former Reagan legal aide who acted as his lawyer, brought suit against the state of California in a highly publicized case in 1981, charging that the teaching of evolution violated his children's religious freedom. The judge found that the guidelines laid down by the Board of Education already emphasized that statements about origins should be phrased conditionally rather than dogmatically, and directed the Board of Education to circulate these guidelines more widely to schools and textbook publishers. Both sides in the dispute claimed victory.

The tack taken by Segraves in the California case was not to request a ban on the teaching of evolution, nor to introduce the Book of Genesis as a scientific text in the classroom. It was, rather, to argue that the "theory of creation," stripped of direct reference to the Bible, has scientific support just as evolution has; and that evolution, also a mere "scientific theory," has been no more proven than creation. The attempt, then, is to set up creation and evolution as equally scientific theories. The same approach is used by Paul Ellwanger, the head of a South Carolina group that drafted a bill which was adopted, in all its essential features, by Arkansas and Louisiana in 1981. These bills mandated that whenever the scientific evidence for evolution was taught in public schools, "balanced treatment" should be given to the "scientific evidences for creation."

The most active, influential, and possibly best-funded creationist group at present is the Institute for Creation Research in San Diego, whose writings I will use to explore the tenets of "scientific creationism." This group includes Henry Morris, its director, who holds a Ph.D. in hydraulic engineering, and Duane Gish, its associate director, who has a Ph.D. in biochemistry. Neither in the creationist literature nor in the scientific literature have I found any reference to professional research by these individuals in genetics, paleontology, taxonomy, anatomy, or any of the other fields most relevant to the study of evolution. Nevertheless, they have written numerous creationist publications on the subject. Morris has stated that evolution is the anti-God conspiracy of Satan[10] and that "the peculiar rings of Saturn, the meteorite swarms

. . . reflect some kind of heavenly catastrophe associated either with Satan's primeval rebellion or his continuing battle with Michael and his angels."[11] He is the author of *The Genesis Flood*, which holds the story of the ark and the global deluge to be literally true.[12] He is also the editor of *Scientific Creationism*,[13] a creationist text available in both a public school edition and a more straightforwardly religious edition. Duane Gish says, "I have always accepted the Bible as God's unchanged and unchangeable revelation to man, and since it describes man and his universe as a special creation of God, I have always been a creationist."[14] The author of *Evolution: The Fossils Say No!*,[15] Gish is a tireless, glib speaker who has debated some eminent evolutionists with considerable success. The voluminous writings of Gish and the rest of the ICR staff are published primarily by Creation-Life Publishers (CLP) in San Diego, from whom you can also order other authors' tracts such as "God's Plan for Air," "God's Plan for Insects," "Unhappy Gays," and "I'm a Woman by God's Design."

Morris and Gish, who have made a career of creationism, speak for millions of devout Christians who view the teaching of evolution as an attack on religion. There are creationist citizens' groups in almost every state, lobbying school boards and legislatures under such names as FLAG ("Family, Life, America under God"), the group that lobbied for passage of the 1981 Arkansas statute that mandated "equal time" for "creation science." This statute was struck down less than ten months after it was passed, by United States District Court Judge William Overton's decision in *McLean* v. *Arkansas Board of Education*.[16] Overton concluded that "creation science" fails to meet the essential characteristics of science, and that creationists, unlike scientists, "do not take data, weigh it against the opposing scientific data, and thereafter reach the conclusions" that species have been individually created. Quoting Henry Morris's assertion that God's written word "is our textbook on the science of Creation," Judge Overton remarked, "While anybody is free to approach a scientific inquiry in any fashion they choose, they cannot properly describe the methodology used as scientific, if they start with a conclusion and refuse to change it regardless of the evidence developed during the course of the investigation." Noting that, like the backers of the bill, the state senator who sponsored the act was "motivated solely by his religious beliefs and desire to see the Biblical version of creation taught in the public schools," the judge concluded that this "unprecedented intrusion in the school curriculum . . . was passed with the specific purpose by the General Assembly of advancing religion," and so violated the First Amendment separation of church and state. In the wake of Judge Overton's decision, however, creationist groups soon affirmed their intent to draft more carefully worded bills. There are no grounds for complacency. Creationism will not soon die.

What is the controversy all about? First of all, almost all scientists hold that the universe is about 14 billion years old, and that the earth and the rest of the solar system were formed about 4½ billion years ago. Life began on earth more than 3 billion years ago, they maintain, and the first very simple living things developed from inanimate matter, through natural chemical and physical processes. All creatures that have ever lived—bacteria, viruses, plants, and animals—are descended from these first forms of life (almost certainly from a single ancestral form). The characteristics of species became modified in time, and single species gave rise to several or many new species by a process of splitting. As a result, the history of living things can be likened to a tree that puts out new twigs and branches even as old ones die and drop off (for most of the species that have ever lived have become extinct). The transformations that each species undergoes often result in better adaptation to the environment, and sometimes in new, unprecedented ways of life. Thus all species, including humans, are descended with modification from common ancestors. The causes of modification and diversification are believed to be entirely natural processes, including factors such as mutation of the genetic material, isolation, and natural selection.

This hypothesis of evolution, of course, cannot be reconciled with a literal reading of the book of Genesis. Genesis says (1:1) that "in the beginning God created the heavens and the earth." He then said, "Let there be light," and He called the light Day, and the darkness He called Night. On the second day He made the firmament and called the firmament Heaven. On the third day He gathered the waters under the heavens together (1:9) and called the dry land Earth. "And God said, 'Let the earth put forth vegetation, plants yielding seed, and fruit trees bearing fruit in which is their seed, each according to its kind, upon the earth.' And it was so" (1:11). On the fourth day, *after* light, day, night, and plants had been made, "God made the two great lights, the greater light to rule the day, and the lesser light to rule the night; he made the stars also" (1:16). On the fifth day He "created the great sea monsters and every living creature that moves, with which the waters swarm, according to their kinds, and every winged bird according to its kind" (1:21). Then (1:25) "God made the beasts of the earth according to their kinds and the cattle according to their kinds, and everything that creeps upon the ground according to its kind. And God saw that it was good."

Finally, on the sixth day, "God created man in his own image, in the image of God He created him; male and female He created them. And God blessed them, and God said to them, 'Be fruitful and multiply, and fill the earth and subdue it; and have dominion over the fish of the sea and over the birds of the air and over every living thing that moves upon the earth,' " (1:27–28).

Aside from the creation of everything by divine fiat in six days, a critical point to note in this story is that each living creature was made "according to its kind"; this is the text that is taken to mean that each "kind" of organism was individually created. Notice also (1:28) that humans are to have dominion over the earth, which is taken to mean that other creatures are made for human benefit.

The order of creation, then, is the heavens and the earth, light, firmament, dry land, plants, sun, moon, and stars, marine animals and birds, terrestrial beasts, and man and woman (together). The sequence in Chapter 2 of Genesis, which tells another creation story, is quite different. "In the day that the Lord God made the earth and the heavens, when no plant of the field was yet in the earth and no herb of the field had yet sprung up . . . then the Lord God formed man of dust from the ground, and breathed into his nostrils the breath of life; and man became a living being" (2:4–7). The Lord then planted a garden with every tree that is pleasant to the sight and good for food (2:9). "Then the Lord God said, 'It is not good that the man should be alone; I will make him a helper fit for him.' So out of the ground the Lord God formed every beast of the field and every bird of the air . . ." (2:18–19). But because the man still did not find a fit helper, the Lord "caused a deep sleep to fall upon the man, and while he slept took one of his ribs and closed up its place with flesh; and the rib which the Lord God had taken from the man He made into a woman, and brought her to the man" (2:21–22).

Nine generations and 1,056 years later, according to Genesis, Noah was born, who had three sons when he was 500 years old (5:32). Because the wickedness of man was great in the earth, God determined to blot out man and beast and creeping things and birds of the air (6:7). But because Noah was righteous and walked with God, the Lord directed Noah to make an ark 300 cubits long, 50 cubits broad, and 30 cubits high, " 'and you shall come into the ark, you, your sons, your wife, and your sons' wives with you. And of every living thing of all flesh, you shall bring two of every sort into the ark, to keep them alive with you; they shall be male and female. . . . Also take with you every sort of food that is eaten and store it up; and it shall serve as food for you and for them.' Noah did this; he did all that God commanded him" (6:18–22). Then the windows of the heavens were opened, rain fell for forty days and forty nights, "and the waters prevailed so mightily upon the earth that all the high mountains under the whole heaven were covered" and "all flesh died that moved upon the earth" (7:19, 21).

Then the waters abated, the ark came to rest upon the mountains of Ararat, and the inhabitants of the ark went forth to be fruitful and multiply. The families of Noah's sons went forth, "and from these the nations spread abroad on the earth after the flood" (10:32), and "the whole earth had one language and few words" (11:1). But the men who settled at Babel in the land of Shinar determined to build a city and a tower with its tops in the heavens. "And the Lord said, 'Behold, they are one people, and they have all one language; and this is only the beginning of what they will do; and nothing that they propose to do will now be impossible for them. Come, let us go down, and there confuse their language, that they may not understand one another's speech" (11:6–7). And so, with their languages confused, they were scattered over the face of all the earth.

I have, of course, left out many of the highlights of Genesis, and noted only those parts of the story that are the source of the creationists' "scientific" explanation of the origin of the earth, living things, humans, and languages. It is undeniably a beautiful parable, the tradition of an ancient nonscientific people as penned, so Biblical scholars say, by no fewer than four writers over the course of centuries. But it was not meant to be, and could not be, a literal history as are, say, the books of Samuel or Ezra; it was a parable meant to affirm to the Israelites that they were the chosen people of God, who had cared for them and delivered them from time immemorial. Like the creation myths of hundreds of other peoples, it was a story wrought from tradition and human experience to provide meaning to human life in an age that could not conceive of science.

Anyone who believes in Genesis as a literal description of history must hold a world view that is entirely incompatible with the idea of evolution, not to speak of science itself. The literalist must believe that different forms of life were individually created and did not develop from common ancestors. Not only species but everything in the physical universe originated not by material, natural processes but by supernatural acts—miracles. Where science insists on material, mechanistic causes that can be understood by physics and chemistry, the literal believer in Genesis invokes unknowable supernatural forces.

Perhaps most importantly, if the world and its creatures developed purely by material, physical forces, it could not have been designed and has no purpose or goal. The fundamentalist, in contrast, believes that everything in the world, every species and every characteristic of every species, was designed by an intelligent, purposeful artificer, and that it was made for a purpose. Nowhere does this contrast apply with more force than to the human species. Some shrink from the conclusion that the human species was not designed, has no purpose, and is the product of mere material mechanisms—but this seems to be the message of evolution.

It is little wonder, then, that the devout fundamentalist recoils from evolution, and that the creationists of today, "scientific" creationists included, rush to defend their religious beliefs. The Bible is their defense. The head of the Institute for Creation Research, Henry Morris, is quite explicit in his book *Studies in The Bible and Science:* "If man wishes to know anything about Creation (the time of Creation, the duration of Creation, the order of Creation, the methods of Creation, or anything else), his sole source of true information is that of divine revelation. God was there when it happened. We were not there. . . . Therefore, we are completely limited to what God has seen fit to tell us, and this information is His written Word. This is our textbook on the science of Creation!"[17] To circumvent the legal strictures against bringing religious doctrine into public schools, though, Morris urges that explicitly theological and Biblical aspects of creationism be taught in the churches, and that school curricula be designed to avoid explicit reference to the Bible. "As long as no religious instruction is given (for example, an exposition of the creation chapters in Genesis), there is no legal problem involved."[18]

What do the creationists believe should be taught? Some people are "theistic evolutionists," who believe that evolution, as conceived and documented by biologists, is the method God has used to achieve his aims. But fundamentalist creationists reject the concept of theistic evolution, which they find theologically repugnant; the Creator of whom they conceive could not have used such cruel, wasteful processes as natural selection and extinction to achieve his ends. Only the Genesis story or something very much like it fits the fundamentalists' concept of creation, because it turns out, according to the creationist literature, that religious views of origins that are not based on a literal interpretation of Genesis are actually evolution in disguise. "There are only *two* world views, evolution and creation. Each of these has many variants. Hinduism and Buddhism are variants of the typical evolutionary world view, beginning as they do with an eternally self-existing universe (the same is true of Confucianism, Taoism, and all the other ancient pagan pantheistic religions)."[19] Their "scientific" theory of creationism entails a personal, omnipotent, intelligent, purposeful Designer—the Creator as traditionally conceived in Judeo-Christian religion. It is the particular concept of creation that the "scientific creationists" espouse that I shall deal with when I show that biology provides no evidence for omnipotence, intelligence, purpose, or design.

The creation "model," as it is described in the Institute for Creation Research publication *Scientific Creationism* (Public School Edition) presupposes a God, or Creator, who created things in the beginning (p. 4). "Once the creation was finished, these processes of *creation* were replaced by processes of *conservation*, which were designed by the Creator to sustain and maintain the basic systems He had created. In addition to the primary concept of a completed creation followed by conservation, the creation model proposes a basic principle of disintegration now at work in nature (since any significant change in a *perfect* primeval creation must be in the direction of imperfection)."[20] We are not told how we can know that the primeval creation was perfect, nor why there should exist a principle of disintegration—although it is clearly necessary if the creationists are to "explain" processes such as mutation and extinction. Belief in an external, omnipotent, Creator God is justified by "the scientific law of *cause-and-effect*. This law, which is universally accepted and followed in every field of science, relates every phenomenon as an effect to a cause. No effect is ever quantitatively 'greater' nor qualitatively 'superior' to its cause. An effect can be lower than its cause but never higher.

"Using causal reasoning, the theistic creationist notes that:

The First Cause of limitless Space	must be infinite
The First Cause of endless Time	must be external
The First Cause of boundless Energy	must be omnipotent
The First Cause of universal Interrelationships	must be omnipresent
The First Cause of Infinite Complexity	must be omniscient
The First Cause of Moral Values	must be moral
The First Cause of Spiritual Values	must be spiritual
The First Cause of Human Responsibility	must be volitional
The First Cause of Human Integrity	must be truthful
The First Cause of Human Love	must be loving
The First Cause of Life	must be living"[21]

Given the deduced nature of the Creator, it follows that "the creation model does include, quite explicitly, the concept of purpose. The

Creator was purposive, not capricious or indifferent, as He planned and then created the universe, with its particles and molecules, its laws and principles, its stars and galaxies, its plants and animals, and finally its human inhabitants."[22] Moreover, "the earth was created specifically to serve as man's home."[23] "In the creationist concept, man is the highest of all creatures, and thus all other created systems must in some way be oriented man-ward, as far as purposes are concerned."[24]

The scientific creationists believe that the earth (and the physical universe) is very young. "In attempting to determine the real age of the earth, it should always be remembered, of course, that recorded history began only several thousand years ago. Not even uranium dating [of rocks] is capable of experimental verification, since no one could actually watch uranium decaying for millions of years to see what happens."[25] Thus, the creation model must interpret the geological record "in terms of essentially continuous deposition, all accomplished in a relatively short time—not instantaneously, of course, but over a period of months or years, rather than millions of years. In effect, this means that the organisms represented in the fossil record must all have been living contemporaneously, rather than scattered in separate time-frames over hundreds of millions of years."[26] To explain the fossil record, then, they say that past processes operated in ways or at rates that are not commensurate with present processes. "Visualize, then, a great hydraulic cataclysm bursting upon the present world, with currents of waters pouring perpetually from the skies and erupting continuously from the earth's crust, all over the world, for weeks on end, until the entire globe was submerged, accompanied by outpourings of magma from the mantle, gigantic earth movements, landslides, tsunamis, and explosions."[27] This model "predicts" that marine invertebrates would be found at the bottom of the geological column since they live on the sea bottom, that fishes would be above them since they live at higher elevations and could also escape burial longer, that mammals and birds would be found in still higher sediments because of their habitat and their mobility, and that the land vertebrates "would tend to be found segregated vertically in the column in order of size and complexity, because of the greater ability of the larger, more diversified animals to escape burial for longer periods of time."[28]

Some humans, of course, survived. "In the creation model, the various tribes and languages [of today] all stemmed from one ancestral population that had developed from a remnant that survived the world-wide flood, which is an integral part of the creation-cataclysm model of earth history. They had been forced to break into a number of small sub-populations by the Creator's direct creative restructuring of their common language into many languages."[29] The model therefore predicts that "the origin of civilization would be located somewhere in the Middle East, near the site of Mount Ararat (where historical tradition indicates the survivors of the antediluvian population emerged from the great cataclysm) or near Babylon (where tradition indicates the confusion of languages took place). This region is located near the geographical center of the post-cataclysm land areas and so would be the natural location for the Creator of mankind, who had providentially preserved a remnant through the Flood, to arrange for the post-diluvian dispersion to begin."[30] "However, the purpose of the book [*Scientific Creationism*] is not necessarily to convince either the teacher or the student that one should believe in creation and reject evolution, or that one should accept Christianity or any other religion."[31]

The creationists reject the notion that simple forms of life could have arisen from nonliving material, for "it seems beyond all question that such complex systems as the DNA molecule could never arise by chance, no matter how big the universe nor how long time. The creation model faces this fact realistically and postulates a great Creator, by whom came life."[32] They admit that species can undergo limited genetic change, because the Creator endowed each of them with a limited amount of genetic variability: "Since the Creator had a purpose for each kind of organism created, He would institute a system which would not only assure its genetic integrity but would also enable it to survive in nature. The genetic system would be such as to maintain its identity as a specific kind while, at the same time, allowing it to adjust its characteristics (within limits) to changes in environment. Otherwise, even very slight changes in its habitat, food supply, etc., might cause its extinction."[33]

Having thus protected themselves against the protest that genetic modifications of species have often been observed to occur, they deny that any "real" novelties can arise in evolution. "Normal variations operate only within the range specified by the DNA for the particular type of organism, so that no truly novel characteristics, producing higher degrees of order or complexity, can appear. Variation is horizontal, not vertical!"[34] The adaptations of species that were bestowed on them by the Creator may be modified slightly by natural selection, but natural processes cannot transform one "kind" of organism into a different "kind." To understand what the creationists mean by a "kind," we must turn to Gish's book *Evolution: The Fossils Say No!*, where he tells us that "a basic animal or plant kind would include all animals or plants which were truly derived from a single stock. In present-day terms, it would be said that they have shared a common gene pool. All humans, for example, are within a single basic kind, *Homo sapiens*. In this case, the basic kind is a single species."[35] "We cannot always be sure, however, what constitutes a separate kind,"[36] for sometimes a "kind" may include a number of similar species.

Why have the creationists gone to such lengths to fit the evidence of geology and biology to the story of Genesis? In part, of course, it is simply to permit them to affirm the literal truth of the Bible. More profoundly, it is because evolution threatens the foundations of the creationist's view of the world. *Scientific Creationism* teaches that creationism should be taught because "creationism is consistent with the innate thoughts and daily experiences of the child and thus is conducive to his mental health . . . it is highly unnatural for him to be told to think of these systems [such as the human body or the ecology of a forest] as chance products of irrational processes."[37] Teaching evolution, in contrast, "is believed by creationists to be harmful to the child or teen-ager since it contradicts his innate consciousness of reality and thus tends to create mental and emotional conflicts within him. It tends to remove all moral and ethical restraints from the student and leads to an animalistic amoralism in practice. It may tend to rob life of meaning and purpose in view of the implanted concept that the student is merely a chance product of a meaningless, random process. Evolutionary philosophy often leads to a conviction that might makes right, leading either to anarchism (uncontrolled evolution) or collectivism (controlled evolution)."[38]

It seems that Darwin is responsible for all the evils of the modern world. Henry Morris, for example, finds in evolution "the seeds of evolutionary racism [which] came to fullest fruition in the form of National Socialism in Germany. The philosopher Friedrich Nietzsche, a contemporary of Charles Darwin and an ardent evolutionist, popularized in Germany his concept of the superman, and then the master race. The ultimate outcome was Hitler, who elevated this philosophy to the status of a national policy."[39] In an exercise of righteous rage, Judge Braswell Dean of Atlanta, speaking perhaps for a fundamentalist multitude, thunders that "this monkey mythology of Darwin is the cause of permissiveness, promiscuity, pills, prophylactics, perversions, abortions, pornography, pollution, poisoning, and proliferation of crimes of all types."[40] The anti-evolution movement is a protest, as Duane Gish says, against "this cancer of evolution-oriented secular humanism that is destroying the minds and faith of our young people."[41]

The fact is, in a scientific sense, there can be no evidence for supernatural special creation. Belief in special creation must rest on faith, on the authority of the Bible and its most literal interpreters. The fundamental conflict, then, is between two incompatible ways to knowledge. Science emphasizes evidence and logical deduction, and is forever uncertain. It deals not with irrefutable facts engraved on stone tablets, but with hypotheses that may be refuted by tomorrow's experiments and concepts formulated by fallible human minds. The best scientific education encourages skepticism, questioning, independent thought, and the use of reason.

But rationalism, skepticism, and science are a threat to respect for traditional authority. The opponents of evolution are part of a larger movement to uphold tradition, authority, and unquestioned values. For example, the *New York Times* reports[42] that in Plano, Texas, "teachers no longer ask students their opinions because to do so, they have been told, is to deny absolute right and wrong." The teacher's word is supposed to be right. Teachers throughout the country, the *Times* reports, are afraid to use controversial materials for fear of bringing down the wrath of parents' groups that are backed by the Moral Majority and other ultraconservative organizations. "I won't use things where a kid has to make a judgment," says one teacher. The move to ban books from classrooms and school libraries has never been stronger: "Since last November there have been attempts to remove, restrict, or deny access to 148 different books in 34 states" in areas as diverse as health, science, social studies, and English.[43] In 1979, 300 communities reported pressure from citizen groups to censor books used in the schools; in 1980, there were 1,200 such incidents.[44]

The fundamentalist notion of what education should be is nicely described by *The New Yorker*'s portrayal of Liberty Baptist College, which is run by Jerry Falwell, head of the Moral Majority.[45] In this college, "education . . . is not a moral and intellectual quest that involves struggle and uncertainty. It is simply the process of learning, or teaching, the right answers. The idea that an individual should collect evidence and decide for himself is anathema. Last spring, Falwell told his congregation that to read anything but the Bible and certain prescribed works of interpretation was at best a waste of time. What bothers the most pious members of his congregation is not just that the schools teach the wrong answers; it is that the schools do not protect children from information that might call their beliefs into question."

The larger question, then, is simply whether education will train people to use their minds or to fall into step with authoritarian doctrine.

Whether creation or evolution will be taught is only this larger question in microcosm; and, in the last analysis, this larger question is a political issue. The political and social consequences of raising a whole country to accept authority without question are terrifying.

These attitudes continue a tradition of anti-intellectualism that has been a powerful force in the United States almost since its foundation. As Richard Hofstadter traces the history in *Anti-Intellectualism in American Life*,[46] a book that has never been more timely than it is today, intellectuals, scientists, and educated people have been distrusted in America ever since John Quincy Adams left the presidency. In a nation of frontiers in which education was almost lacking, a nation that viewed intellectuals as an elite contrary to the spirit of populism, a nation that prided itself on repudiating the institutions of European civilization, intellectual activity fell into low repute. The life of the mind was considered, and is still widely considered, to be opposed to emotion, manly character, practicality, and democracy. Intellectuals were considered responsible for social changes that led ordinary people to be uprooted from their familiar comforts by forces they did not understand.

This ground was fertile for the growth of fundamentalist religion, which involved the replacement of the intellectual tradition of Judaism and European Christianity with a religion of pure emotion, of direct personal access to God. It is in this context that what Hofstadter calls the "one hundred percent mentality" flourished; that is, "a mind totally committed to the full range of the dominant popular fatuities and determined that no one shall have a right to challenge them. . . . The one-hundred percenter . . . will tolerate no ambiguities, no equivocations, no reservations, and no criticism." The Scopes trial of 1925 was, as Hofstadter says, the perfect dramatization of everything that was at stake in the confrontation between the fundamentalist and the modern mind. The anti-evolution laws of Tennessee and other states were ultimately an attempt to prevent children from rejecting the values and beliefs of their parents.

The opponents of evolution who call themselves "creationists" are, almost without exception, fundamentalist Christians. It is important, however, to realize that they do not represent religion as a whole. Religion is not necessarily opposed to evolution. The only official *(ex cathedra)* statement of the Roman Catholic Church,[47] for example, has been the declaration in 1950 by Pope Pius XII that "the teaching of the Church leaves the doctrine of evolution an open question." I well remember, as a student in a Catholic high school in 1959, the enthusiasm of some of my Jesuit teachers for Teilhard de Chardin's *The Phenomenon of Man,* an attempt to synthesize evolution with Catholic dogma. More recently, the editor of the Jesuit magazine *America* has agreed that "the idea of evolution . . . by now has the status not of a mere hypothesis but of a scientific fact."[48]

At a 1972 hearing on whether or not the California science curriculum should include the Biblical story of creation, many religious leaders argued against mixing religion into science. For example, the Episcopal Reverend Julian Barrett said that "if at any time, *any* theological doctrine should be proven incorrect under the impact of scientific *knowledge,* I shall discard that theological doctrine. . . . In *true* science, every thesis, every theory, every so-called 'law' is 'on the line,' subject to continued testing before our ever-expanding body of knowledge. . . . That Biblical myth-story [of creation] was but one of many such which were developed by primitive religions. Over 100 years ago modern science began to dismantle the superstructure of religious myth-stories of origins, and of the Genesis story in particular, by means of scientific investigation. In doing so, science rendered Biblical religion an inestimable service in that religion was enabled to recover a simple truth about the Book of Genesis: i.e., that it was a religious and therefore theological document and not a scientific treatise."[49] At the same hearing, Rabbi Amiel Wohl, of the Congregation B'nai Israel in Sacramento, remarked that "in our Jewish religious traditions today, we find in that account [Genesis] great moral power, eloquence and beauty. . . . But, as for the teaching of science, we would never purport to place the Creation Epic as a scientific theory of creation. We understand it as a theological statement. . . . The truth of Adam and Eve stories, or of any other Biblical tales, does not rise or fall on their scientific demonstrability, but rather on their moral and symbolic teaching."

The opponents of evolution, then, are not the religious leaders who understand Genesis to contain symbolic truth, but the fundamentalists who are incapable of recognizing metaphor, and insist on interpreting Genesis literally. They hold a Manichean, black-or-white, all-or-none view of the world, in which any indication that science is imprecise, uncertain, or mistaken is jumped on as "proof" that creation is true. Their Biblical analysis is as absurd as are their scientific views; they cannot admit of metaphor, parable, or historical scholarship in their reading of the Bible; for to admit that one passage should be read metaphorically is to admit ambiguity in the whole, which would bring into doubt the passages that serve as authority for their uncompromising moral and social positions.

Why are the creationists successful? In part, because they offer simple, unambiguous, and certain answers to those who cannot live with complexity, ambiguity, uncertainty. In part, because scientists, whose lives are in their laboratories, are usually reluctant to commit themselves to public education and public debate. Another reason is that it is often hard to understand scientific issues without mastering a specialized vocabulary and technical details, and most people react to these by turning off. The creationists have the advantage of riding roughshod over all intricacies and details. They present an appealingly black-and-white caricature of science that makes few demands on its audience's intellect.

The sociologist Dorothy Nelkin,[50] in her explanation of the creationists' success, finds other reasons for the enormous gap in understanding between scientists and non-scientists. Most people, she points out, believe that science is a collection of immutable facts, so that scientists are expected to provide exact, definitive answers. When their answers are provisional or tentative, they are not considered scientific—despite the fact that all scientists regard their work as provisional and tentative. Non-scientists also tend to think that the validity of a scientific theory depends on its implications—whether it is socially "safe" and emotionally appealing—rather than on scientific verification. In addition, Nelkin points out, science does not provide a sense of personal integration. It is, from a human point of view, remote and heartless in its conclusions. As a result, many people turn from emotionally unsatisfying, seemingly incomprehensible technical theories to a fundamentalist world view that satisfies their personal convictions and emotional needs. Thus, Nelkin says, for many non-scientists "beliefs need no evidence."

That is the central focus of this book. I believe that if creationism, for which there can be no evidence, is promoted to the status of science, the message to future generations will be clear: beliefs need no evidence.

Evolutionary biology is founded on evidence, though this often doesn't come across in high school or college courses in general biology. Textbooks (in every subject) have a way of feeding students information as if it shouldn't be questioned, instead of challenging the student to discover why scientists hold the beliefs they do. Evolution shouldn't be accepted on faith any more than creation should, but few books explain the evidence for evolution. In exploring the choice between evolution and creation, it will be necessary to understand some fairly technical arguments about genes and fossils; and it will be necessary, too, to go into the philosophical foundations of knowledge. The battles in the courtroom in the coming years are going to turn on the questions "What is a theory? What is the difference between theory and fact? What can science prove?" I hope this book will convey not only the evidence for evolution, but the nature of the scientific adventure that is under attack.

TWO

―――――

THE GROWTH OF EVOLUTIONARY SCIENCE

Today, the theory of evolution is an accepted fact for everyone but a fundamentalist minority, whose objections are based not on reasoning but on doctrinaire adherence to religious principles.
　　—James D. WATSON, 1965*

In 1615, Galileo was summoned before the Inquisition in Rome. The guardians of the faith had found that his "proposition that the sun is the center [of the solar system] and does not revolve about the earth is foolish, absurd, false in theology, and heretical, because expressly contrary to Holy Scripture." In the next century, John Wesley declared that "before the sin of Adam there were no agitations within the bowels of the earth, no violent convulsions, no concussions of the earth, no earthquakes, but all was unmoved as the pillars of heaven." Until the seventeenth century, fossils were interpreted as "stones of a peculiar sort, hidden by the Author of Nature for his own pleasure." Later they

―――――

* James D. Watson, a molecular biologist, shared the Nobel Prize for his work in discovering the structure of DNA.

were seen as remnants of the Biblical deluge. In the middle of the eighteenth century, the great French naturalist Buffon speculated on the possibility of cosmic and organic evolution and was forced by the clergy to recant: "I abandon everything in my book respecting the formation of the earth, and generally all of which may be contrary to the narrative of Moses." For had not St. Augustine written, "Nothing is to be accepted save on the authority of Scripture, since greater is that authority than all the powers of the human mind"?

When Darwin published *The Origin of Species*, it was predictably met by a chorus of theological protest. Darwin's theory, said Bishop Wilberforce, "contradicts the revealed relations of creation to its Creator." "If the Darwinian theory is true," wrote another clergyman, "Genesis is a lie, the whole framework of the book of life falls to pieces, and the revelation of God to man, as we Christians know it, is a delusion and a snare." When *The Descent of Man* appeared, Pope Pius IX was moved to write that Darwinism is "a system which is so repugnant at once to history, to the tradition of all peoples, to exact science, to observed facts, and even to Reason herself, [that it] would seem to need no refutation, did not alienation from God and the leaning toward materialism, due to depravity, eagerly seek a support in all this tissue of fables."[1] Twentieth-century creationism continues this battle of medieval theology against science.

One of the most pervasive concepts in medieval and post-medieval thought was the "great chain of being," or *scala naturae*.[2] Minerals, plants, and animals, according to this concept, formed a gradation, from the lowliest and most material to the most complex and spiritual, ending in man, who links the animal series to the world of intelligence and spirit. This "scale of nature" was the manifestation of God's infinite benevolence. In his goodness, he had conferred existence on all beings of which he could conceive, and so created a complete chain of being, in which there were no gaps. All his creatures must have been created at once, and none could ever cease to exist, for then the perfection of his divine plan would have been violated. Alexander Pope expressed the concept best:

> Vast chain of being! which from God began,
> Natures aethereal, human, angel, man,
> Beast, bird, fish, insect, what no eye can see,
> No glass can reach; from Infinite to thee,
> From thee to nothing.—On superior pow'rs
> Were we to press, inferior might on ours;

Or in the full creation leave a void,
Where, one step broken, the great scale's destroy'd;
From Nature's chain whatever link you strike,
Tenth, or ten thousandth, breaks the chain alike.

Coexisting with this notion that all of which God could conceive existed so as to complete his creation was the idea that all things existed for man. As the philosopher Francis Bacon put it, "Man, if we look to final causes, may be regarded as the centre of the world . . . for the whole world works together in the service of man . . . all things seem to be going about man's business and not their own."

"Final causes" was another fundamental concept of medieval and post-medieval thought. Aristotle had distinguished final causes from efficient causes, and the Western world saw no reason to doubt the reality of both. The "efficient cause" of an event is the mechanism responsible for its occurrence: the cause of a ball's movement on a pool table, for example, is the impact of the cue or another ball. The "final cause," however, is the goal, or purpose for its occurrence: the pool ball moves because I wish it to go into the corner pocket. In post-medieval thought there was a final cause—a purpose—for everything; but purpose implies intention, or foreknowledge, by an intellect. Thus the existence of the world, and of all the creatures in it, had a purpose; and that purpose was God's design. This was self-evident, since it was possible to look about the world and see the palpable evidence of God's design everywhere. The heavenly bodies moved in harmonious orbits, evincing the intelligence and harmony of the divine mind; the adaptations of animals and plants to their habitats likewise reflected the divine intelligence, which had fitted all creatures perfectly for their roles in the harmonious economy of nature.

Before the rise of science, then, the causes of events were sought not in natural mechanisms but in the purposes they were meant to serve, and order in nature was evidence of divine intelligence. Since St. Ambrose had declared that "Moses opened his mouth and poured forth what God had said to him," the Bible was seen as the literal word of God, and according to St. Thomas Aquinas, "Nothing was made by God, after the six days of creation, absolutely new." Taking Genesis literally, Archbishop Ussher was able to calculate that the earth was created in 4004 B.C. The earth and the heavens were immutable, changeless. As John Ray put it in 1701 in *The Wisdom of God Manifested in the Works of the Creation*, all living and nonliving things were "created by God at first, and by Him conserved to this

Day in the same State and Condition in which they were first made."[3]

The evolutionary challenge to this view began in astronomy. Tycho Brahe found that the heavens were not immutable when a new star appeared in the constellation Cassiopeia in 1572. Copernicus displaced the earth from the center of the universe, and Galileo found that the perfect heavenly bodies weren't so perfect: the sun had spots that changed from time to time, and the moon had craters that strongly implied alterations of its surface. Galileo, and after him Buffon, Kant, and many others, concluded that change was natural to all things.

A flood of mechanistic thinking ensued. Descartes, Kant, and Buffon concluded that the causes of natural phenomena should be sought in natural laws. By 1755, Kant was arguing that the laws of matter in motion discovered by Newton and other physicists were sufficient to explain natural order. Gravitation, for example, could aggregate chaotically dispersed matter into stars and planets. These would join with one another until the only ones left were those that cycled in orbits far enough from each other to resist gravitational collapse. Thus order might arise from natural processes rather than from the direct intervention of a supernatural mind. The "argument from design"—the claim that natural order is evidence of a designer—had been directly challenged. So had the universal belief in final causes. If the arrangement of the planets could arise merely by the laws of Newtonian physics, if the planets could be born, as Buffon suggested, by a collision between a comet and the sun, then they did not exist for any purpose. They merely came into being through impersonal physical forces.

From the mutability of the heavens, it was a short step to the mutability of the earth, for which the evidence was far more direct. Earthquakes and volcanoes showed how unstable terra firma really is. Sedimentary rocks showed that materials eroded from mountains could be compacted over the ages. Fossils of marine shells on mountaintops proved that the land must once have been under the sea. As early as 1718, the Abbé Moro and the French academician Bernard de Fontenelle had concluded that the Biblical deluge could not explain the fossilized oyster beds and tropical plants that were found in France. And what of the great, unbroken chain of being if the rocks were full of extinct species?

To explain the facts of geology, some authors—the "catastrophists" —supposed that the earth had gone through a series of great floods and other catastrophes that successively extinguished different groups of animals. Only this, they felt, could account for the discovery that higher and lower geological strata had different fossils. Buffon, however, held that to explain nature we should look to the natural causes we see operating around us: the gradual action of erosion and the slow buildup of land during volcanic eruptions. Buffon thus proposed what came to be the foundation of geology, and indeed of all science, the principle of uniformitarianism, which holds that the same causes that operate now have always operated. By 1795, the Scottish geologist James Hutton had suggested that "in examining things present we have data from which to reason with regard to what has been." His conclusion was that since "rest exists not anywhere," and the forces that change the face of the earth move with ponderous slowness, the mountains and canyons of the world must have come into existence over countless aeons.

If the entire nonliving world was in constant turmoil, could it not be that living things themselves changed? Buffon came close to saying so. He realized that the earth had seen the extinction of countless species, and supposed that those that perished had been the weaker ones. He recognized that domestication and the forces of the environment could modify the variability of many species. And he even mused, in 1766, that species might have developed from common ancestors:

> If it were admitted that the ass is of the family of the horse, and different from the horse only because it has varied from the original form, one could equally well say that the ape is of the family of man, that he is a degenerate man, that man and ape have a common origin; that, in fact, all the families among plants as well as animals have come from a single stock, and that all animals are descended from a single animal, from which have sprung in the course of time, as a result of process or of degeneration, all the other races of animals. For if it were once shown that we are justified in establishing these families; if it were granted among animals and plants there has been (I do not say several species) but even a single one, which has been produced in the course of direct descent from another species . . . then there would no longer be any limit to the power of nature, and we should not be wrong in supposing that, with sufficient time, she has been able from a single being to derive all the other organized beings.[4]

This, however, was too heretical a thought; and in any case, Buffon thought the weight of evidence was against common descent. No new species had been observed to arise within recorded history, Buffon wrote; the sterility of hybrids between species appeared an impossible barrier to such a conclusion; and if species had emerged gradually, there should have been innumerable intermediate variations between the horse and ass, or any other species. So Buffon concluded: "But this [idea of a common ancestor] is by no means a proper representation of nature. We are assured by the authority of revelation that all animals have participated equally in the grace of direct Creation and that the first pair of every species issued fully formed from the hands of the Creator."

Buffon's friend and protégé, Jean Baptiste de Monet, the Chevalier de Lamarck, was the first scientist to take the big step. It is not clear what led Lamarck to his uncompromising belief in evolution; perhaps it was his studies of fossil molluscs, which he came to believe were the ancestors of similar species living today. Whatever the explanation, from 1800 on he developed the notion that fossils were not evidence of extinct species but of ones that had gradually been transformed into living species. To be sure, he wrote, "an enormous time and wide variation in successive conditions must doubtless have been required to enable nature to bring the organization of animals to that degree of complexity and development in which we see it at its perfection"; but "time has no limits and can be drawn upon to any extent."

Lamarck believed that various lineages of animals and plants arose by a continual process of spontaneous generation from inanimate matter, and were transformed from very simple to more complex forms by an innate natural tendency toward complexity caused by "powers conferred by the supreme author of all things." Various specialized adaptations of species are consequences of the fact that animals must always change in response to the needs imposed on them by a continually changing environment. When the needs of a species change, so does its behavior. The animal then uses certain organs more frequently than before, and these organs, in turn, become more highly developed by such use, or else "by virtue of the operations of their own inner senses." The classic example of Lamarckism is the giraffe: by straining upward for foliage, it was thought, the animal had acquired a longer neck, which was then inherited by its offspring.

In the nineteenth century it was widely believed that "acquired" characteristics—alterations brought about by use or disuse, or by the direct influence of the environment—could be inherited. Thus it was perfectly reasonable for Lamarck to base his theory of evolutionary change partly on this idea. Indeed, Darwin also allowed for this possibility, and the inheritance of acquired characteristics was not finally proven impossible until the 1890s.

Lamarck's ideas had a wide influence; but in the end did not convince many scientists of the reality of evolution. In France, Georges

Cuvier, the foremost paleontologist and anatomist of his time, was an influential opponent of evolution. He rejected Lamarck's notion of the spontaneous generation of life, found it inconceivable that changes in behavior could produce the exquisite adaptations that almost every species shows, and emphasized that in both the fossil record and among living animals there were numerous "gaps" rather than intermediate forms between species. In England, the philosophy of "natural theology" held sway in science, and the best-known naturalists continued to believe firmly that the features of animals and plants were evidence of God's design. These devout Christians included the foremost geologist of the day, Charles Lyell, whose *Principles of Geology* established uniformitarianism once and for all as a guiding principle. But Lyell was such a thorough uniformitarian that he believed in a steady-state world, a world that was always in balance between forces such as erosion and mountain building, and so was forever the same. There was no room for evolution, with its concept of steady change, in Lyell's world view, though he nonetheless had an enormous impact on evolutionary thought, through his influence on Charles Darwin.

Darwin (1809–1882) himself, unquestionably one of the greatest scientists of all time, came only slowly to an evolutionary position. The son of a successful physician, he showed little interest in the life of the mind in his early years. After unsuccessfully studying medicine at Edinburgh, he was sent to Cambridge to prepare for the ministry, but he had only a half-hearted interest in his studies and spent most of his time hunting, collecting beetles, and becoming an accomplished amateur naturalist. Though he received his B.A. in 1831, his future was quite uncertain until, in December of that year, he was enlisted as a naturalist aboard *H.M.S. Beagle,* with his father's very reluctant agreement. For five years (from December 27, 1831, to October 2, 1836) the *Beagle* carried him about the world, chiefly along the coast of South America, which it was the *Beagle*'s mission to survey. For five years Darwin collected geological and biological specimens, made geological observations, absorbed Lyell's *Principles of Geology*, took voluminous notes, and speculated about everything from geology to anthropology. He sent such massive collections of specimens back to England that by the time he returned he had already gained a substantial reputation as a naturalist.

Shortly after his return, Darwin married and settled into an estate at Down where he remained, hardly traveling even to London, for the rest of his life. Despite continual ill health, he pursued an extraordinary range of biological studies: classifying barnacles, breeding pigeons, experimenting with plant growth, and much more. He wrote no fewer than sixteen books and many papers, read voraciously, corresponded extensively with everyone, from pigeon breeders to the most eminent scientists, whose ideas or information might bear on his theories, and kept detailed notes on an amazing variety of subjects. Few people have written authoritatively on so many different topics: his books include not only *The Voyage of the Beagle, The Origin of Species,* and *The Descent of Man,* but also *The Structure and Distribution of Coral Reefs* (containing a novel theory of the formation of coral atolls which is still regarded as correct), *A Monograph on the Sub-class Cirripedia* (the definitive study of barnacle classification), *The Various Contrivances by Which Orchids are Fertilised by Insects, The Variation of Animals and Plants Under Domestication* (an exhaustive summary of information on variation, so crucial to his evolutionary theory), *The Effects of Cross and Self Fertilisation in the Vegetable Kingdom* (an analysis of sexual reproduction and the sterility of hybrids between species), *The Expression of the Emotions in Man and Animals* (on the evolution of human behavior from animal behavior), and *The Formation of Vegetable Mould Through the Action of Worms.* There is every reason to believe that almost all these books bear, in one way or another, on the principles and ideas that were inherent in Darwin's theory of evolution. The worm book, for example, is devoted to showing how great the impact of a seemingly trivial process like worm burrowing may be on ecology and geology if it persists for a long time. The idea of such cumulative slight effects is, of course, inherent in Darwin's view of evolution: successive slight modifications of a species, if continued long enough, can transform it radically.

When Darwin embarked on his voyage, he was a devout Christian who did not doubt the literal truth of the Bible, and did not believe in evolution any more than did Lyell and the other English scientists he had met or whose books he had read. By the time he returned to England in 1836 he had made numerous observations that would later convince him of evolution. It seems likely, however, that the idea itself did not occur to him until the spring of 1837, when the ornithologist John Gould, who was working on some of Darwin's collections, pointed out to him that each of the Galápagos Islands, off the coast of Ecuador, had a different kind of mockingbird. It was quite unclear whether they were different varieties of the same species, or different species. From this, Darwin quickly realized that species are not the discrete, clear-cut entities everyone seemed to imagine. The possibility of transformation entered his mind, and it applied to more than the mockingbirds: "When comparing . . . the birds from the separate islands of the Galápagos archipelago, both with one another and with those from the American mainland, I was much struck how entirely vague and arbitrary is the distinction between species and varieties."

In July 1837 he began his first notebook on the "Transmutation of Species." He later said that the Galápagos species and the similarity between South American fossils and living species were at the origin of all his views.

> During the voyage of the *Beagle* I had been deeply impressed by discovering in the Pampean formation great fossil animals covered with armour like that on the existing armadillos; secondly, by the manner in which closely allied animals replace one another in proceeding southward over the continent; and thirdly, by the South American character of most of the productions of the Galápagos archipelago, and more especially by the manner in which they differ slightly on each island of the group; none of these islands appearing to be very ancient in a geological sense. It was evident that such facts as these, as well as many others, could be explained on the supposition that species gradually become modified; and the subject has haunted me.

The first great step in Darwin's thought was the realization that evolution had occurred. The second was his brilliant insight into the possible cause of evolutionary change. Lamarck's theory of "felt needs" had not been convincing. A better one was required. It came on September 18, 1838, when after grappling with the problem for fifteen months, "I happened to read for amusement Malthus on Population, and being well prepared to appreciate the struggle for existence which everywhere goes on from long-continued observation of the habits of animals and plants, it at once struck me that under these circumstances favorable variations would tend to be preserved, and unfavorable ones to be destroyed. The result of this would be the formation of new species. Here, then, I had at last got a theory by which to work."

Malthus, an economist, had developed the pessimistic thesis that the exponential growth of human populations must inevitably lead to famine, unless it were checked by war, disease, or "moral restraint." This emphasis on exponential population growth was apparently the catalyst for Darwin, who then realized that since most natural populations of animals and plants remain fairly stable in numbers, many more individuals are born than survive. Because individuals vary in their characteristics, the struggle to survive must favor some variant individuals over others. These survivors would then pass on their characteristics to fu-

ORIGINAL
ANCESTOR

FIGURE 1. Some species of Galápagos finches. Several of the most different species are represented here; intermediate species also exist. Clockwise from lower left are a male ground-finch (the plumage of the female resembles that of the tree-finches); the vegetarian tree-finch; the insectivorous tree-finch; the warbler-finch; and the woodpecker-finch, which uses a cactus spine to extricate insects from crevices. The slight differences among these species, and among species in other groups of Galápagos animals such as giant tortoises, were one of the observations that led Darwin to formulate his hypothesis of evolution. (From D. Lack, *Darwin's Finches* [Oxford: Oxford University Press, 1944].)

ture generations. Repetition of this process generation after generation would gradually transform the species.

Darwin clearly knew that he could not afford to publish a rash speculation on so important a subject without developing the best possible case. The world of science was not hospitable to speculation, and besides, Darwin was dealing with a highly volatile issue. Not only was he affirming that evolution had occurred, he was proposing a purely material explanation for it, one that demolished the argument from design in a single thrust. Instead of publishing his theory, he patiently amassed a mountain of evidence, and finally, in 1844, collected his thoughts in an essay on natural selection. But he still didn't publish. Not until 1856, almost twenty years after he became an evolutionist, did he begin what he planned to be a massive work on the subject, tentatively titled *Natural Selection*.

Then, in June 1858, the unthinkable happened. Alfred Russel Wallace (1823–1913), a young naturalist who had traveled in the Amazon Basin and in the Malay Archipelago, had also become interested in evolution. Like Darwin, he was struck by the fact that "the most closely allied species are found in the same locality or in closely adjoining localities and . . . therefore the natural sequence of the species by affinity is also geographical." In the throes of a malarial fever in Malaya, Wallace conceived of the same idea of natural selection as Darwin had, and sent Darwin a manuscript "On the Tendency of Varieties to Depart Indefinitely from the Original Type." Darwin's friends Charles Lyell and Joseph Hooker, a botanist, rushed in to help Darwin establish the priority of his ideas, and on July 1, 1858, they presented to the Linnean Society of London both Wallace's paper and extracts from Darwin's 1844 essay. Darwin abandoned his big book on natural selection and condensed the argument into a 490-page "abstract" that was published on November 24, 1859, under the title *The Origin of Species by Means of Natural Selection; or, the Preservation of Favored Races in the Struggle for Life*. Because it was an abstract, he had to leave out many of the detailed observations and references to the literature that he had amassed, but these were later provided in his other books, many of which are voluminous expansions on the contents of *The Origin of Species*.

The first five chapters of the *Origin* lay out the theory that Darwin had conceived. He shows that both domesticated and wild species are variable, that much of that variation is hereditary, and that breeders, by conscious selection of desirable varieties, can develop breeds of pigeons, dogs, and other forms that are more different from each other than species or even families of wild animals and plants are from each other.

The differences between related species then are no more than an exaggerated form of the kinds of variations one can find in a single species; indeed, it is often extremely difficult to tell if natural populations are distinct species or merely well-marked varieties.

Darwin then shows that in nature there is competition, predation, and a struggle for life.

> Owing to this struggle, variations, however slight and from whatever cause proceeding, if they be in any degree profitable to the individuals of a species, in their infinitely complex relations to other organic beings and to their physical conditions of life, will tend to the preservation of such individuals, and will generally be inherited by the offspring. The offspring, also, will thus have a better chance of surviving, for, of the many individuals of any species which are periodically born, but a small number can survive. I have called this principle, by which each slight variation, if useful, is preserved, by the term natural selection, in order to mark its relation to man's power of selection.

Darwin goes on to give examples of how even slight variations promote survival, and argues that when populations are exposed to different conditions, different variations will be favored, so that the descendants of a species become diversified in structure, and each ancestral species can give rise to several new ones. Although "it is probable that each form remains for long periods unaltered," successive evolutionary modifications will ultimately alter the different species so greatly that they will be classified as different genera, families, or orders.

Competition between species will impel them to become more different, for "the more diversified the descendants from any one species become in structure, constitution and habits, by so much will they be better enabled to seize on many and widely diversified places in the polity of nature, and so be enabled to increase in numbers." Thus different adaptations arise, and "the ultimate result is that each creature tends to become more and more improved in relation to its conditions. This improvement inevitably leads to the greater advancement of the organization of the greater number of living beings throughout the world." But lowly organisms continue to persist, for "natural selection, or the survival of the fittest, does not necessarily include progressive development—it only takes advantage of such variations as arise and are beneficial to each creature under its complex relations of life." Probably no organism has reached a peak of perfection, and many lowly forms of life continue to exist, for "in some cases variations or individual differences of a favorable nature may never have arisen for natural selection to act on or accumulate. In no case, probably, has time sufficed for the utmost possible amount of development. In some few cases there has been what we must call retrogression of organization. But the main cause lies in the fact that under very simple conditions of life a high organization would be of no service. . . ."

In the rest of *The Origin of Species*, Darwin considers all the objections that might be raised against his theory; discusses the evolution of a great array of phenomena—hybrid sterility, the slave-making instinct

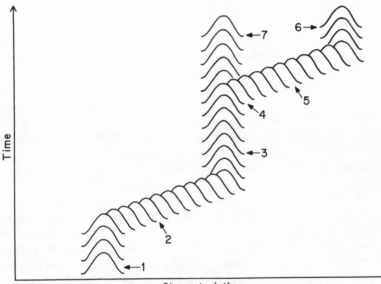

FIGURE 2. Processes of evolutionary change. A characteristic that is variable (1) often shows a bell-shaped distribution—individuals vary on either side of the average. Evolutionary change (2) consists of a shift in successive generations, after which the characteristic may reach a new equilibrium (3). When the species splits into two different species (4), one of the species may undergo further evolutionary change (5) and reach a new equilibrium (6). The other may remain unchanged (7) or not. Each population usually remains variable throughout this process, but the average is shifted, ordinarily by natural selection.

of ants, the similarity of vertebrate embryos; and presents an enormous body of evidence for evolution. He draws his evidence from comparative anatomy, embryology, behavior, geographic variation, the geographic distribution of species, the study of rudimentary organs, atavistic variations ("throwbacks"), and the geological record to show how all of biology provides testimony that species have descended with modification from common ancestors.

Darwin's triumph was in synthesizing ideas and information in ways that no one had quite imagined before. From Lyell and the geologists he learned uniformitarianism: the cause of past events must be found in natural forces that operate today; and these, in the vastness of time, can accomplish great change. From Malthus and the nineteenth-century economists he learned of competition and the struggle for existence. From his work on barnacles, his travels, and his knowledge of domesticated varieties he learned that species do not have immutable essences but are variable in all their properties and blend into one another gradually. From his familiarity with the works of Whewell, Herschel, and other philosophers of science he developed a powerful method of pursuing science, the "hypothetico-deductive" method, which consists of formulating a hypothesis or speculation, deducing the logical predictions that must follow from the hypothesis, and then testing the hypothesis by seeing whether or not the predictions are verified. This was by no means the prevalent philosophy of science in Darwin's time.[5]

Darwin brought biology out of the Middle Ages. For divine design and unknowable supernatural forces he substituted natural material causes that could be studied by the methods of science. Instead of catastrophes unknown to physical science he invoked forces that could be studied in anyone's laboratory or garden. He replaced a young, static world by one in which there had been constant change for countless aeons. He established that life had a history, and this proved the essential view that differentiated evolutionary thought from all that had gone before.

For the British naturalist John Ray, writing in 1701, organisms had no history—they were the same at that moment, and lived in the same places, doing the same things, as when they were first created. For Darwin, organisms spoke of historical change. If there has indeed been such a history, then fossils in the oldest rocks must differ from those in younger rocks: trilobites, dinosaurs, and mammoths will not be mixed together but will appear in some temporal sequence. If species come from common ancestors, they will have the same characteristics, modified for different functions: the same bones used by bats for flying will be used by horses for running. If species come from ancestors that lived in different environments, they will carry the evidence of their history with them in the form of similar patterns of embryonic development and in vestigial, rudimentary organs that no longer serve any function. If species have a history, their geographical distribution will reflect it: oceanic islands won't have elephants because they wouldn't have been able to get there.

Once the earth and its living inhabitants are seen as the products of historical change, the theological philosophy embodied in the great chain of being ceases to make sense; the plenitude, or fullness, of the world becomes not an eternal manifestation of God's bountiful creativity but an illusion. For most of earth's history, most of the present species have not existed; and many of those that did exist do so no longer. But the scientific challenge to medieval philosophy goes even deeper. If evolution has occurred, and if it has proceeded from the natural causes that Darwin envisioned, then the adaptations of organisms to their environment, the intricate construction of the bird's wing and the orchid's flower, are evidence not of divine design but of the struggle for existence. Moreover, and this may be the deepest implication of all, Darwin brought to biology, as his predecessors had brought to astronomy and geology, the sufficiency of efficient causes. No longer was there any reason to look for final causes or goals. To the questions "What purpose does this species serve? Why did God make tapeworms?" the answer is "To no purpose." Tapeworms were not put here to serve a purpose, nor were planets, nor plants, nor people. They came into existence not by design but by the action of impersonal natural laws.

By providing materialistic, mechanistic explanations, instead of miraculous ones, for the characteristics of plants and animals, Darwin brought biology out of the realm of theology and into the realm of science. For miraculous spiritual forces fall outside the province of science; all of science is the study of material causation.

Of course, *The Origin of Species* didn't convince everyone immediately. Evolution and its material cause, natural selection, evoked strong protests from ecclesiastical circles, and even from scientists.[6] The eminent geologist Adam Sedgwick, for example, wrote in 1860 that species must come into existence by creation,

a power I cannot imitate or comprehend; but in which I can believe, by a legitimate conclusion of sound reason drawn from the laws and harmonies of Nature. For I can see in all around me a design and purpose, and a mutual adaptation of parts which I *can* comprehend, and which prove that there is exterior to, and above, the mere phenomena of Nature a great prescient and designing cause. . . . The pretended physical philosophy of modern days strips man of all his moral attributes, or holds them of no account in the estimate of his origin and place in the created world. A cold atheistical materialism is the tendency of the so-called material philosophy of the present day.

Among the more scientific objections were those posed by the French paleontologist François Pictet, and they were echoed by many others. Since Darwin supposes that species change gradually over the course of thousands of generations, then, asked Pictet, "Why don't we find these gradations in the fossil record . . . and why, instead of collecting thousands of identical individuals, do we not find more intermediary forms? . . . How is it that the most ancient fossil beds are rich in a variety of diverse forms of life, instead of the few early types Darwin's theory leads us to expect? How is it that no species has been seen to evolve during human history, and that the 4000 years which separates us from the mummies of Egypt have been insufficient to modify the crocodile and the ibis?" Pictet protested that, although slight variations might in time alter a species slightly, "all known facts demonstrate . . . that the prolonged influence of modifying causes has an action which is constantly restrained within sufficiently confined limits."

The anatomist Richard Owen likewise denied "that . . . variability is progressive and unlimited, so as, in the course of generations, to change the species, the genus, the order, or the class." The paleontologist Louis Agassiz insisted that organisms fall into discrete groups, based on uniquely different created plans, between which no intermediates could exist. He chose the birds as a group that showed the sharpest of boundaries. Only a few years later, in 1868, the fossil *Archaeopteryx,* an exquisite intermediate between birds and reptiles, demolished Agassiz's argument, and he had no more to say on the unique character of the birds.

Within twelve years of *The Origin of Species,* the evidence for evolution had been so thoroughly accepted that the philosopher and mathematician Chauncey Wright could point out that among the students of science, "orthodoxy has been won over to the doctrine of evolution." However, Wright continued, "While the general doctrine of evolution has thus been successfully redeemed from theological condemnation, this is not yet true of the subordinate hypothesis of Natural Selection."

Natural selection turned out to be an extraordinarily difficult concept for people to grasp. St. George Mivart, a Catholic scholar and scientist, was not unusual in equating natural selection with chance. "The theory of Natural Selection may (though it need not) be taken in such a way as to lead man to regard the present organic world as formed, so to speak, *accidentally*, beautiful and wonderful as is the confessedly haphazard result." Many like him simply refused to understand that natural selection is the antithesis of chance and consequently could not see how selection might cause adaptation or any kind of progressive evolutionary change. Even in the 1940s there were those, especially among paleontologists, who felt that the progressive evolution of groups like the horses, as revealed by the fossil record, must have had some unknown cause other than natural selection. Paradoxically, then, Darwin had convinced the scientific world of evolution where his predecessors had failed; but he had not convinced all biologists of his truly original theory, the theory of natural selection.

Natural selection fell into particular disrepute in the early part of the twentieth century because of the rise of genetics—which, as it happened, eventually became the foundation of the modern theory of evolution. Darwin's supposition that variation was unlimited, and so in time could give rise to strikingly different organisms, was not entirely convincing because he had no good idea of where variation came from. In 1865, the Austrian monk Gregor Mendel discovered, from his crosses of pea plants, that discretely different characteristics such as wrinkled versus smooth seeds were inherited from generation to generation without being altered, as if they were caused by particles that passed from parent to offspring. Mendel's work was ignored for thirty-five years, until, in 1900, three biologists discovered his paper and realized that it held the key to the mystery of heredity. One of the three, Hugo de Vries, set about to explore the problem as Mendel had, and in the course of his studies of evening primroses observed strikingly different variations arise, *de novo*. The new forms were so different that de Vries believed they represented new species, which had arisen in a single step by alteration or, as he called it, mutation, of the hereditary material.

In the next few decades, geneticists working with a great variety of organisms observed many other drastic changes arise by mutation: fruit flies *(Drosophila)*, for example, with white instead of red eyes or curled instead of straight wings. These laboratory geneticists, especially Thomas Hunt Morgan, an outstanding geneticist at Columbia University, asserted that evolution must proceed by major mutational steps, and that mutation, not natural selection, was the cause of evolution. In their eyes, Darwin's theory was dead on two counts: evolution was not gradual, and it was not caused by natural selection. Meanwhile, naturalists, taxonomists, and breeders of domesticated plants and animals continued to believe in Darwinism, because they saw that populations and species differed quantitatively and gradually rather than in big jumps, that most variation was continuous (like height in humans) rather than discrete, and that domesticated species could be altered by artificial selection from continuous variation.

The bitter conflict between the Mendelian geneticists and the Darwinians was resolved in the 1930s in a "New Synthesis" that brought the opposing views into a "neo-Darwinian" theory of evolution.[7] Slight variations in height, wing length, and other characteristics proved, under careful genetic analysis, to be inherited as particles, in the same way as the discrete variations studied by the Mendelians. Thus a large animal simply has inherited more particles, or genes, for large size than a smaller member of the species has. The Mendelians were simply studying particularly well marked variations, while the naturalists were studying more subtle ones. Variations could be very slight, or fairly pronounced, or very substantial, but all were inherited in the same manner. All these variations, it was shown, arose by a process of mutation of the genes.

Three mathematical theoreticians, Ronald Fisher and J. B. S. Haldane in England and Sewall Wright in the United States, proved that a newly mutated gene would not automatically form a new species. Nor would it automatically replace the preexisting form of the gene, and so transform the species. Replacement of one gene by a mutant form of the gene, they said, could happen in two ways. The mutation could enable its possessors to survive or reproduce more effectively than the old form; if so, it would increase by natural selection, just as Darwin had said. The new characteristic that evolved in this way would ordinarily be considered an improved adaptation.

Sewall Wright pointed out, however, that not all genetic changes in species need be adaptive. A new mutation might be no better or worse than the preexisting gene—it might simply be "neutral." In small populations such a mutation could replace the previous gene purely by chance—a process he called random genetic drift. The idea, put crudely, is this. Suppose there is a small population of land snails in a cow pasture, and that 5 percent of them are brown and the rest are yellow. Purely by chance, a greater percentage of yellow snails than of brown ones get crushed by cows' hooves in one generation. The snails breed, and there will now be a slightly greater percentage of yellow snails in the next generation than there had been. But in the next generation, the yellow ones may suffer more trampling, purely by chance. The proportion of yellow offspring will then be lower again. These random events cause fluctuations in the percentage of the two types. Wright proved mathematically that eventually, if no other factors intervene, these fluctuations will bring the population either to 100 percent yellow or 100 percent brown, purely by chance. The population will have evolved, then, but not by natural selection; and there is no improvement of adaptation.

During the period of the New Synthesis, though, genetic drift was emphasized less than natural selection, for which abundant evidence was discovered. Sergei Chetverikov in Russia, and later Theodosius Dobzhansky working in the United States, showed that wild populations of fruit flies contained an immense amount of genetic variation, including the same kinds of mutations that the geneticists had found arising in their laboratories. Dobzhansky and other workers went on to show that these variations affected survival and reproduction: that natural selection was a reality. They showed, moreover, that the genetic differences among related species were indeed compounded of the same kinds of slight genetic variations that they found within species. Thus the taxonomists and the geneticists converged onto a neo-Darwinian theory of evolution: evolution is due not to mutation *or* natural selection, but to both. Random mutations provide abundant genetic variation; natural selection, the antithesis of randomness, sorts out the useful from the deleterious, and transforms the species.

In the following two decades, the paleontologist George Gaylord Simpson showed that this theory was completely adequate to explain the fossil record, and the ornithologists Bernhard Rensch and Ernst Mayr, the botanist G. Ledyard Stebbins, and many other taxonomists showed that the similarities and differences among living species could be fully explained by neo-Darwinism. They also clarified the meaning of "species." Organisms belong to different species if they do not interbreed when the opportunity presents itself, thus remaining genetically distinct. An ancestral species splits into two descendant species when different populations of the ancestor, living in different geographic regions, become so genetically different from each other that they will not or cannot interbreed when they have the chance to do so. As a result, evolution can happen without the formation of new species: a single species can be genetically transformed without splitting into several descendants. Conversely, new species can be formed without much genetic change. If one population becomes different from the rest of its species in, for example, its mating behavior, it will not interbreed with the other populations. Thus it has become a new species, even though it may be identical to its "sister species" in every respect except its behavior. Such a new species is free to follow a new path of genetic change, since it does not become homogenized with its sister species by interbreeding. With time, therefore, it can diverge and develop different adaptations.

The conflict between the geneticists and the Darwinians that was resolved in the New Synthesis was the last major conflict in evolutionary science. Since that time, an enormous amount of research has confirmed most of the major conclusions of neo-Darwinism. We now know that populations contain very extensive genetic variation that continually arises by mutation of preexisting genes. We also know what genes are and how they become mutated. Many instances of the reality of natural selection in wild populations have been documented, and there is extensive evidence that many species form by the divergence of different populations of an ancestral species.

The major questions in evolutionary biology now tend to be of the form, "All right, factors x and y both operate in evolution, but how important is x compared to y?" For example, studies of biochemical genetic variation have raised the possibility that nonadaptive, random change (genetic drift) may be the major reason for many biochemical differences among species. How important, then, is genetic drift compared to natural selection? Another major question has to do with rates of evolution: Do species usually diverge very slowly, as Darwin thought, or does evolution consist mostly of rapid spurts, interspersed with long periods of constancy? Still another question is raised by mutations, which range all the way from gross changes of the kind Morgan studied to very slight alterations. Does evolution consist entirely of the substitution of mutations that have very slight effects, or are major mutations sometimes important too? Partisans on each side of all these questions argue vigorously for their interpretation of the evidence, but they don't doubt that the major factors of evolution are known. They simply emphasize one factor or another. Minor battles of precisely this kind go on continually in every field of science; without them there would be very little advancement in our knowledge.

Within a decade or two of *The Origin of Species*, the belief that living organisms had evolved over the ages was firmly entrenched in biology.

As of 1982, the historical existence of evolution is viewed as fact by almost all biologists. To explain how the fact of evolution has been brought about, a theory of evolutionary mechanisms—mutation, natural selection, genetic drift, and isolation—has been developed.[8] But exactly what is the evidence for the fact of evolution?

THREE

THE LEGACY
OF THE
TAXONOMISTS

There are many generalizations in biology, but precious few theories. Among these, the theory of evolution is by far the most important, because it draws together from the most varied sources a mass of observations which would otherwise remain isolated; it unites all the disciplines concerned with living beings; it establishes order among the extraordinary variety of organisms and closely binds them to the rest of the earth; in short, it provides a causal explanation of the living world and its heterogeneity.
—FRANÇOIS JACOB, 1973*

When Darwin began his work, biology was a descriptive science. Its major question was "what," rather than "how" or "why." Generations of naturalists had been more concerned to document what kinds of plants and animals existed than to probe the problems of how they

---

* François Jacob, a geneticist, shared the Nobel Prize for his role in discovering mechanisms by which genes are expressed in development.

lived or how they had come into being. The chief preoccupation of biology was taxonomy—describing species and assigning them to categories. More than one biologist saw his work as a holy office, dedicated to the glorification of the Creator. The father of modern taxonomy, Linnaeus, opened his *Systema Naturae* in 1757 with the ejaculation, "O Jehova, Quam ampla sunt Tua Opera!"—"How bountiful are Thy works!" His book was a catalogue of the known species of animals, a systematization of knowledge undertaken to celebrate the creation.

Linnaeus's great work consisted of giving each species, or "kind," of animal and plant a name such as *Sciurus vulgaris*, the common European squirrel, and *Sciurus volans*, the flying squirrel. These species were filed into categories within categories: the squirrels being combined with rats and mice into the rodent order, within the class Mammalia. Later taxonomists expanded the levels of categories, so that we now have phyla embracing classes, classes embracing orders, and so on down the taxonomic hierarchy: orders, families, genera, species. However, the existence of a filing system implies some criteria for filing, and this raises a serious problem. Why should any one classification be better than any other? Why should the whale be classified with the mammals because of its lungs, rather than among the fishes because of its fins?

Linnaeus classified species together on the basis of gross overall similarity—and sometimes on what must have been a very arbitrary basis. Plants were classified by the number of stamens in the flower; bats were classified with primates because of the position of the female's breasts. Subsequent generations of taxonomists devoted themselves to probing anatomy more deeply, and found anatomical similarities between bats and shrews, and between horses and rhinoceroses, which they used to reclassify them.

The great anatomist Richard Owen, who became one of Darwin's chief critics, proposed that certain similarities between different organisms were "homologous"—that the skeletal similarity of the bat's wing and the horse's leg revealed them to be the "same" structure serving different functions. Other similarities—the fins and streamlined forms of whales and fishes, for example, were claimed to be "analogous"— superficially similar structures with different underlying anatomies or modes of embryonic development. Organisms, then, should be classified together if they had homologous structures. But what did homology mean? In the pre-Darwinian world of natural history, it could only mean that they were constructed by God on the same plan: that God, for reasons of economy, had used one blueprint for the structure of vertebrates, another for insects, and so on, modifying each blueprint to

the needs of the individual species he had created. The aim of classification, then, was to understand and reveal God's plan. A "natural" classification was to be, in a sense, a map of God's mind.

To Lamarck and Darwin, homology came to have a very different
meaning. If the bat and the horse had descended from a common ancestor, it made perfect sense that the one had become adapted for flying
and the other for running by different modifications of the same ancestral structures. This conclusion instantly gave a scientific, rather than
theological, meaning and rationale for classification. A natural classification would group those species that were related by descent, and separate those that were merely similar. If a scientific reason can be
established for classifying whales among the mammals, then the "analogous" features of whales and fish must be the consequence of "convergent" evolution—similar structures independently evolved from
different antecedents in response to similar environmental conditions.
If we classify species by mere similarity, there can be dozens of classifications, depending on what characteristics you look at, none of which has
a claim to be better than the others. However, there can be only one true
evolutionary history, hence only one true basis for an evolutionary
classification, if evolution is true.

The hypothesis of evolution solves at a single stroke hundreds of
puzzles that had troubled anatomists and other comparative biologists.
It explains, for example, why similar functions are often served by
modifications of different characteristics. To manipulate the bamboo on
which it feeds, the giant panda uses a thumblike modification of a wrist
bone, instead of the finger that in primates is modified into a thumb.[1]
The creationists are at a loss to explain why God's blueprint should be
different for the panda and for primates, but the difference is an immediate consequence of a fundamental principle of evolution: that natural
selection acts on whatever variations arise fortuitously. Similarly, why
should the poinsettia attract pollinators by bright red leaves instead of
petals, which serve this function in most plants? Why doesn't the poinsettia have petals? The answer is clear when we observe that plants
related to the poinsettia also lack petals: if the ancestor of all these plants
had lost its petals, the poinsettia would have had to make do with
modifications of the structures left to it. If God had equipped very
different organisms for similar ways of life, there is no reason why He
should not have provided them with identical structures, but in fact the
similarities are always superficial. The dog and the Tasmanian wolf, as
Darwin pointed out, have similar feeding habits; but the Tasmanian
wolf, like opossums and other marsupials, has three premolars and

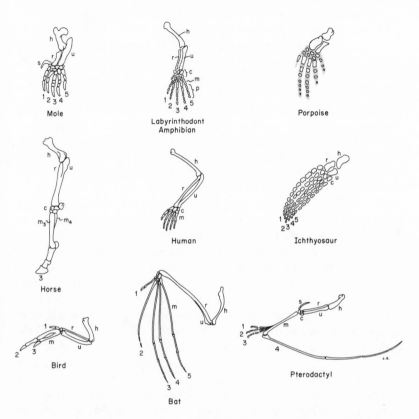

FIGURE 3. Homologous bones of the left forelimbs of various tetrapod
vertebrates. The extinct labyrinthodont amphibian illustrates the
ancestral condition. The major changes are in the loss of bones
(especially in the horse and bird), the additional segments in the
fingers of the extinct ichthyosaur (a marine reptile), the fusion of
certain elements (especially in the horse), and the changes in the
shapes of different bones in different species. The homologous elements are humerus (*h*), radius (*r*), ulna (*u*), carpals (*c*), metacarpals
(*m*), and digits 1 through 5. Nonhomologous bones developed from
cartilage are marked *s*. The stippled area in the porpoise hand
represents cartilage. Although the hands of the porpoise and the
ichthyosaur have both been modified for swimming, and those of
the bird, bat, and pterodactyl were all modified for flight, the
modifications have been different in each case.

47

four molars, while the dog has four premolars and only two molars.

The facts of embryology, the study of development, also make little sense except in the light of evolution. Why should species that ultimately develop adaptations for utterly different ways of life be nearly indistinguishable in their early stages? How does God's plan for humans and sharks require them to have almost identical embryos? Why should terrestrial salamanders, if they were not descended from aquatic ancestors, go through a larval stage entirely within the egg, with gills and fins that are never used, and then lose these features before they hatch?

In the 1920s, embryology passed out of its descriptive phase into the realm of experimentation, with results that would have pleased Darwin. Take, for example, the experiments of the French embryologist A. Hampé.[2] The drumstick of a chicken contains two bones—the thin, short fibula and the larger tibiotarsus. Reptiles have a well-developed tibia (shank bone) and fibula, and separate ankle bones, which anatomists have long believed correspond to part of the bird's tibiotarsus. Hampé inserted a thin sheet of mica between the developing tibia and fibula of a chicken embryo, and the reptilian arrangement—separate ankle bones and a well-developed fibula—developed perfectly. Apparently the bird retains the ancient reptilian program of development, but has modified it so that the tibia ordinarily prevents the fibula from growing, and unless the fibula grows down to make contact with the ankle bones, they become fused with the tibia.

An even more striking experiment was reported recently by E. J. Kollar and C. Fisher in the journal *Science*.[3] Even though birds normally never develop teeth, these experimenters were able to induce embryonic tissue from the jaw of a chicken to develop teeth, complete with enamel, by laying it over tissue from the jaw of a mouse embryo. The teeth developed from the chicken cells, but in response to chemical signals from the underlying mouse tissue. It is evident, then, that birds still have the ability to respond to such signals, even though they have not had teeth for more than 100 million years.

An experiment of this kind can only be interpreted to mean that birds possess genetic information for making teeth—or reptilian legs— that is never used, but which has been bequeathed to them by their ancestors. It certainly makes no sense in a creationist framework. In fact, the existence of vestigial structures was one of the major dilemmas that was resolved by the hypothesis of evolution. Why should the Creator have bestowed useless, rudimentary structures on his creatures? "To complete the scheme of nature," said the pre-Darwinians. But, said Darwin, this is inconsistent: "the boa-constrictor has rudiments of hind legs and of a pelvis, and if it be said that these bones have been retained 'to complete the scheme of nature,' why ... have they not been retained by other snakes, which do not possess even a vestige of these same bones?" Darwin drove the point home with example after example: the reduction of one lung in snakes, the teeth of fetal calves which never cut through the gums, the useless wings of flightless beetles that are sealed beneath fused wing covers, the rudimentary pistils in the male flowers of certain plants. "It would be impossible," Darwin said, "to name one of the higher animals in which some part or another is not in a rudimentary condition." Vestigial structures make no sense except in the light of evolutionary history, just as the spelling and usage of so many words in English make no sense except in the light of their Latin or Old Norse antecedents.

Anatomy, embryology, classification—these fields were given a new meaning by evolution, and continued to confirm the theory of descent with modification. So, in similar ways, did other fields that

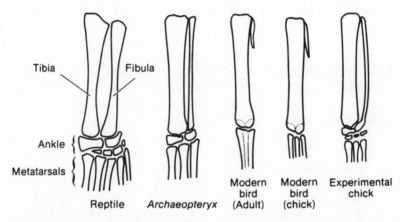

FIGURE 4. Hampé's experiment demonstrating that the development of a bird's leg is a modification of the reptilian process of development. The separate metatarsals of the chick (fourth from left) become fused with one another in the adult bird. The ankle bones are not evident; some are fused with the metatarsals, and others with the tibia. Hampé's experimental chicks (right) developed a long fibula, and the ankle bones and metatarsals developed into separate bones, as is the case in reptiles and in *Archaeopteryx*. (From T. H. Frazzetta, *Complex Adaptations in Evolving Populations* [Sunderland, Mass.: Sinauer Associates, Inc., 1975].)

developed in the post-Darwinian era. Biochemists and physiologists, for example, find a red protein in vertebrates and certain insects. It performs the same function—carrying oxygen—but the detailed structure of the hemoglobin protein, the sequence of amino acids that make up the protein molecule, is entirely different in the two groups. It really isn't at all the same molecule in insects and vertebrates. A creationist might suppose that God would provide the same molecule to serve the same function, but a biologist would never expect evolution to follow exactly the same path twice.

In recent years, molecular biologists have been able to determine the exact sequence of the amino acids that are strung together to make proteins such as hemoglobin or cytochrome-c, a molecule that is important in respiration. It turns out, as Walter Fitch and Emanuel Margoliash discovered,[4] that almost without exception, species which the taxonomists had already judged to be closely related in an evolutionary sense have more similar protein molecules than those that are supposed to be more distantly related. The similarity of the molecules isn't related to whether or not the animals have similar ways of life. It is entirely a matter of their genealogical relationship.

For biologists, one of the benefits of the great age of discovery was that explorers brought back to Europe natural curiosities from the far corners of the earth. Their efforts often had the same motivation as the early taxonomists' attempts at classification: cataloging the wonders of God's creation. Their collections, however, revealed something that finally struck Darwin with full force on his own voyage, and more than any other evidence impelled him to believe in evolution. There were peculiar regularities to the ways in which animals and plants were distributed throughout the world that could only be viewed as capricious if they were the handiwork of a Creator. Why, for example, should Australia be populated by marsupial versions of the wolf, the mole, the squirrel, and the mouse, rather than by the real articles? Why should oceanic islands lack most kinds of animals except for those few whose features suggested an ability to cross great expanses of ocean? Why should there be a woodpecker in the Argentine pampas, and hardly any trees for hundreds of miles? Why, as Darwin found in the Galápagos Islands, should the economy of nature be filled by a host of similar species of finches, each slightly modified for an ecological role that birds like warblers, parrots, and woodpeckers play in South America? In general, it appeared that a given taxonomic group of species was not distributed throughout the world wherever its special habitat occurred, as an economical Creator might have ordained. Rather, similar ecological roles were played by unrelated species in different places; and in any given area, closely related species were adapted to different ways of life. Woodpeckers don't occur in the Galápagos, or on any other oceanic islands, where there are trees; but some do occur in Argentina, where they feed on the ground and yet possess the structures that forest-dwelling woodpeckers use for climbing and drilling into wood.

The only rational—that is, scientific—explanation for such patterns must be that species were not distributed over the face of the earth by the Creator but had originated in different places and had dispersed from there. Thus there are no moles in Australia or woodpeckers in the Galápagos simply because they couldn't get there. The molelike and wolflike animals of Australia are marsupials, clearly related to each other, because only marsupial ancestors had reached Australia. Once such evolution was acknowledged, the field of taxonomy could expand into biogeography, the scientific study of the reasons for the geographical distributions of animals and plants.

As more information on the distribution of organisms came to light, embarrassing cases emerged. For example, evolutionists were hard put to explain why lungfishes and cichlid fishes—freshwater fishes such as the Oscar and the Jack Dempsey in the aquarium trade—occurred only in South America, Africa, and, in the case of lungfishes, Australia, too. Accepting evolution, biologists had to suppose either that the African and South American forms really weren't related to each other, or that they had somehow gotten from one continent to the other. The anatomical similarities were so great that convergent evolution was ruled out, so some biologists began to fantasize that the southern continents had once been connected by narrow bridges of land. For this, geologists could provide no evidence. Others, such as Philip Darlington[5] in his massive treatise on zoogeography in 1957, proposed that these groups had arisen in Asia or Africa, and dispersed by way of the Bering land bridge from eastern Asia to Alaska, down through North America into South America. This isn't as ridiculous as it sounds, because Alaska and Siberia have indeed been connected at various times in the past, and the fossils from that region show that it was once much warmer. Unfortunately for Darlington's hypothesis, however, there was no fossil evidence that cichlids, or any of the many other groups of freshwater fish now limited to the southern continents, ever occurred in North America. The major exception is lungfishes, their fossils being plentiful in ancient North American rocks.

There was, however, another explanation: that the continents haven't always been where they are now. In 1912, Alfred Wegener, a meteorologist, introduced what seemed to be an absurd idea—continental drift. Wegener drew on several lines of evidence to argue that the continents, despite their apparent solidity, have moved about the earth. Continental drift could explain why South America and Africa have complementary shapes, why certain rock formations are found on both sides of the Atlantic, and why eastern South America and western Africa bear evidence of glaciation, even though glaciers cannot arise from the ocean. Despite this and other evidence, most geologists dismissed Wegener's hypothesis for the next fifty years. What, they protested, could possibly cause continents to move?

Only in the late 1950s did a few geologists begin to reconsider the idea. First, there was indisputable evidence, from the position of magnetic particles in rocks, that the continents must have had different orientations to the poles that they have now. More importantly, geologists such as Arthur Holmes of the University of Edinburgh realized that geophysical mechanisms capable of causing continental movement do exist. The continents are made of less dense material than the underlying basalt, and float on it like gobs of congealed fat on cold soup. The basalt itself slowly moves because of the heat of the earth's interior. Thermal convection currents from within the earth bring material up to the surface along a great ridge that runs down the middle of the Atlantic Ocean, where it pushes the earth's crust to either side.[6]

Following the revelation of this mechanism, evidence was quickly amassed by geologists to test what now seemed to be a plausible theory. Within a decade, the geological profession had moved en masse from ridiculing the idea of continental drift to accepting it. We are now quite sure that separate continental masses aggregated shortly before the Age of Reptiles into a single continent that has been named Pangaea. It then fragmented into a northern continent, Laurasia, and a southern landmass, Gondwanaland. By the end of the Age of Reptiles (the Mesozoic era), Laurasia had fragmented into Eurasia and North America. Gondwanaland broke up into South America, Africa, Australia, Antarctica, and India, which moved northward and collided with Asia, forming the Himalayas as it did so. The westward movement of the Americas and the eastward drift of Eurasia is still going on, and causes immense stresses in the earth's crust, so that the rim of the Pacific is ringed with faults and volcanoes.

The importance of continental drift for biology is that it solves all sorts of biogeographical dilemmas. Ancient groups of organisms such as lungfishes are found as fossils throughout the world, as would be expected if they evolved before Pangaea broke up. Groups that are thought, on the basis of fossils, to have arisen somewhat more recently are distributed over the continents that were once joined into either Laurasia or Gondwanaland, but are not worldwide. Thus, cichlids are in South America and Africa because their ancestors once swam the rivers of the continent of Gondwanaland. Other groups, such as the horses, evolved after the continents split up. The fossil history of the horses is almost entirely limited to North America; only in the last few million years did they make their way, presumably by the Alaska-Siberia land bridge, into Asia and Africa. By then, Australia and South America were too isolated for them to reach. Continental drift also explains why Antarctica bears a fossil record of forests, large reptiles, and marsupials that are similar to those on other southern continents. Thus biogeography began as one of the sources of evidence that compelled Darwin to believe in evolution; graduated into an evolutionary science that had to invent unconvincing ad hoc hypotheses to explain its data in evolutionary terms; and was resolved by new geological evidence into a simple, consistent history of life.

One of the many reasons for believing that organisms have a common evolutionary history is that their characteristics are often hierarchically arranged. Since evolution is supposed to proceed by a series of sequential splitting events, a new characteristic that evolves in one particular branch of the tree of life is likely to be passed on to all the descendants of that branch. Within this group, another new characteristic evolves, and is then passed on to descendants of that particular species. Thus, for example, the four-legged condition evolved in amphibians, and is retained by most of their descendants. Among these, the ancestors of the mammals evolved a single-boned lower jaw. Among some of their descendants, the rodents developed gnawing incisors, and so on. There is a nesting of groups within groups, as a consequence of common ancestry. Objects like minerals that are not descended from common ancestors cannot be arranged in this way.

Because of the hierarchical nature of evolution, it is often possible to unravel the genealogy of a group of organisms with fair confidence, even in the absence of a good fossil record. For example, we can postulate that because lizards and most other terrestrial vertebrates have limbs and two lungs, the ancestors of snakes probably had these characteristics. If this is so, then the boa constrictors and pythons represent primitive snakes, because they have two lungs and rudimentary hind limbs. The other snakes, which have one lung and no rudimentary limbs, are then a separate, more derived branch. These species, such as the garter snakes, have large scales on their heads, and they lack fangs or heat-sensitive pits below the nostrils. The vipers, however, have fangs, which we therefore believe arose in one group of highly derived ("advanced") snakes. Within the vipers there is a split between the Old World species that lack heat-sensitive pits and the New World species such as copperheads and rattlesnakes which possess this new feature. Among the New World species, the rattlesnakes possess a new feature, the rattle. Among the rattlesnakes, the "primitive" species such as the pygmy rattlesnake retain the large head scales that almost all other snakes possess, while the diamondback rattlesnake and its relatives have become altered, and possess small head scales. Thus even without a good fossil record, we can describe the genealogy of the snakes.

The major stumbling block in this logic is convergent evolution: if a similar characteristic evolved independently in two groups, they may be mistakenly classified as relatives. Usually, however, convergent evolution can be spotted because it gives rise to contradictory evolutionary trees. For example, flowering plants are divided into two great groups, the monocots (lilies, orchids, and others) and the dicots (oaks, roses, and

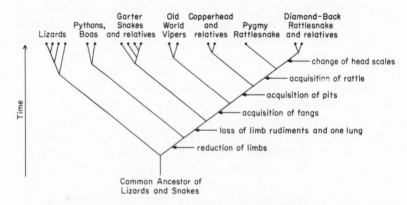

FIGURE 5. A simplified phylogenetic tree of the lizards and snakes. The common ancestor of lizards and snakes, according to the deductive methods used in phylogenetic analysis, had limbs and two lungs, and lacked fangs, heat-sensitive pits, and a rattle.

many others). One major difference between the two groups is in the number of seedling leaves or cotyledons each possesses: monocots have one and dicots two. There are, however, many other differences between the groups in the structure of their stems, flowers, and leaves. The spring beauties, which typically have two cotyledons, are also like other dicots in all these other respects. One species of spring beauty, however, has only a single cotyledon.[7] It would be absurd to think it was closely related to the lilies on this account; for we would then have to suppose that it was convergently similar to other spring beauties, and to other dicotyledonous plants, in dozens or hundreds of other characteristics. The simplest explanation is that it has converged toward the monocots in this one characteristic.

It is possible, then, to deduce phylogeny, that is, genealogical history, by a careful, logical analysis of which organisms share which characteristics. A genealogy derived in this way may be considered a hypothesis, always subject to possible revision. If the hypothesis makes predictions that are borne out, we gain more confidence that it is correct. For example, the entomologists William Brown, Edward Wilson, and Frank Carpenter hypothesized, using the kind of logic I have described, that ants arose from ancestors that resemble certain wasps. They predicted that if a missing link were ever found, it would have certain specified characteristics (including a reduced thorax and an unconstricted abdomen), and that it would be more than 70 million years old. In 1967, several years after this prediction had been made, they acquired a 100-million-year-old fossil ant that matched their prediction in almost every important detail.[8] Another kind of prediction that taxonomists can make on the basis of logical deduction is that certain kinds of characteristics, not yet examined, should fit a phylogenetic tree. Such predictions have been beautifully borne out in many cases when the molecular structure of various species' proteins has been examined. Figure 6 shows, for example, the phylogenetic tree of a wide variety of vertebrates based just on the structure of one protein, cytochrome-c.[9] With a few minor exceptions (in the exact arrangement of a few of the birds, for example), this "chemical" tree perfectly matches the genealogy of these vertebrates as it had been understood for decades before such molecular data became available. Thus a hypothesis about evolutionary history can be corroborated by using entirely independent kinds of data.

The similarities among organisms that the evolutionary biologist attributes either to common ancestry or to convergent evolution are explained by the creationist as separate creations for the same function.

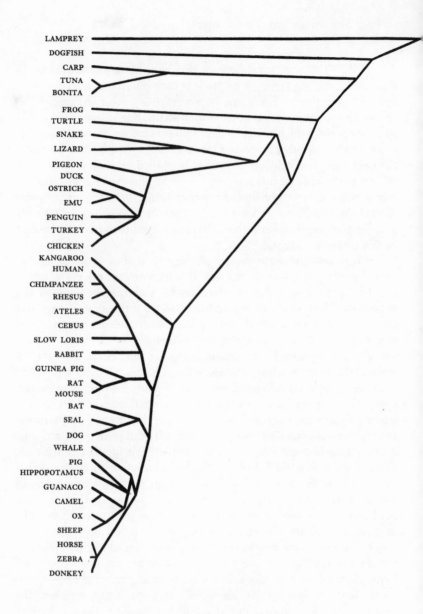

LAMPREY
DOGFISH
CARP
TUNA
BONITA
FROG
TURTLE
SNAKE
LIZARD
PIGEON
DUCK
OSTRICH
EMU
PENGUIN
TURKEY
CHICKEN
KANGAROO
HUMAN
CHIMPANZEE
RHESUS
ATELES
CEBUS
SLOW LORIS
RABBIT
GUINEA PIG
RAT
MOUSE
BAT
SEAL
DOG
WHALE
PIG
HIPPOPOTAMUS
GUANACO
CAMEL
OX
SHEEP
HORSE
ZEBRA
DONKEY

But creationists are quite silent on the question of why the wings of birds and bats should be so utterly different in structure—why the structure of a bat's wing should resemble that of a monkey's hand more than that of a bird's wing. They are quite silent on the question of why the cytochrome-c of bats, which is important in energy production in metabolism, should be similar to that of whales and humans, with utterly different ways of life, and why the bat's protein should be different from that of either flying or flightless birds. The creationist can offer no prediction about which organisms should have similar molecules; the evolutionist can, and has done so.

Phylogenetic trees are often expressed, quite imprecisely, by the classification of organisms. Species are placed in categories within categories: genera within families, families within orders. These categories above the species level are called "higher taxa," and many taxonomists feel that the species that are classified together into a higher taxon should be those that stem from a common ancestor. The higher the taxon, the more remote the common ancestry. Thus all the mammals, class Mammalia, share a remote ancestor; the dogs, cats, bears, and so on (order Carnivora) a more recent ancestor; the various cats (family Felidae) a more recent ancestor still.

Every such classification is "real" in some ways, but arbitrary in others. Referring to Figure 6, for example, it clearly would be improper to classify the kangaroo along with the camel, ox, and sheep in the order Artiodactyla, because this would imply that the kangaroo and the camel share a more recent ancestor than they actually do. On the other hand, it would be quite arbitrary whether we combined the camel, its relative the guanaco, the ox, and the sheep into one family, or split them into two, as is usually done (guanaco and camel in the Camelidae, ox and sheep in the Bovidae). Whether we make one large family or two small ones doesn't violate our representation of them as having a common

FIGURE 6. A phylogenetic tree of the vertebrates, showing the order in which various species diverged from their common ancestors. Each segment of the tree represents an evolving lineage that subsequently split into the separate groups we recognize today. The genealogical relationships represented by this tree are almost identical with those that have long been postulated on the basis of anatomical features, but this diagram is based on independent data, the amino acid sequence of the protein cytochrome-c of each species. (From M. Goodman, in *Prog. Biophys. Mol. Biol.* 38 [1982]; courtesy of Pergamon Press and Morris Goodman.)

ancestor. These species, in fact, are placed along with the pig and hippopotamus in the order Artiodactyla, which reflects their common ancestry.

The very nature of the evolutionary process ensures that the limits we set on each group of related species are arbitrary. Some taxonomists, for example, emphasize the difference between lizards and snakes by placing them in separate orders. Others are more impressed by their similarities, and by the existence of a group of intermediate forms (the so-called "blind" snakes), and combine them all into one order. Many bird taxonomists recognize separate families for the thrushes, wrens, and European flycatchers; but Ernst Mayr and Dean Amadon find so many species that are intermediate between these groups that they combine them into one large family.[10]

The reason for this arbitrariness is, of course, that we are trying to impose discrete categories on organisms that have arisen by gradual divergence. Thus many species are in various intermediate stages and cannot be clearly categorized, any more than we can divide a person's life discretely into adolescence and adulthood.

The fact that the dividing line between thrushes and wrens or lizards and snakes is arbitrary is immensely important. The "scientific creationists" accept that "microevolution"—slight differences between similar, related species—may well have arisen from the slight genetic alterations that biologists argue for. But the creationists claim that "macroevolution"—the origin of radically different "kinds" of animals and plants—is another story. They say that there are unbridgeable gaps between the major families, orders, and classes of animals. Unfortunately for them, such "gaps" aren't unbridgeable and often don't exist at all. There is no gap between thrushes and wrens, between lizards and snakes, or between sharks and skates. A complete gamut of intermediate species runs from the great white shark to the butterfly ray, and each step in the series is a small one, corresponding to the slight differences that separate similar species.

Naturally, some gaps do exist. Few people would have trouble distinguishing reptiles from mammals, for example. But in many such cases, we find that quite discrete categories, if we look just at living species, become more and more blurred as we go back in the fossil record. Horses and rhinoceroses become indistinguishable; hoofed animals and carnivores become less and less distinct as we reach the beginning of the Tertiary period; and mammals become completely indistinguishable from reptiles when we encounter the therapsids, the mammal-like reptiles of the Permian period, 250 million years ago.[11] The gaps we see today among living species exist because of the extinction of intermediate forms. Many gaps so far haven't been filled in by fossils, but these are chiefly in groups that do not fossilize well, or groups such as the major phyla of animals that diverged in the very remote past, for which the fossil record is poor.

According to the creationists,[12] classification would be impossible if evolution had occurred, because then there would be no gaps in the evolutionary continuum. But of course there are gaps, because of extinction; and where fossils or living species fill in the continuum, classification does indeed become most difficult and arbitrary.

Once the phylogenetic relationships among species have been determined, it is possible to make many important generalizations about evolution. One of these is that when species diverge from one another, they do so in some characteristics but not others. Thus the various species of cats differ in size and color, but not in the basic structure of their teeth. Therefore it is wrong to suppose, as many people do, that an "advanced" species will be "advanced" in all respects, and it is equally wrong to think that a "missing link" will be intermediate in every way between the species that it links together. Thus snakes are more primitive* than humans in that they do not have hair or a four-chambered heart, but humans are more primitive in that they have limbs, two lungs rather than one, and many other features. Snakes would no doubt consider us a very lowly form of life if they could conceive of evolution and took their own complicated jaw structure as the criterion of evolutionary advancement.

Every major group of related species is distinguished by one or more characteristics that may be termed "evolutionary novelties"— major features like the feathers of birds or the legs of the terrestrial vertebrates that adapt them to a special way of life. One of the major debates, from Darwin's day to the present, has been the argument over how such major features have evolved. We look at a bird's feathers and

---

* The terms "primitive" and "advanced" have unfortunate connotations. As I explain in this paragraph, snakes are no more primitive than humans, except in the sense that the split between reptiles and mammals preceded the split between primates and other mammals. "Primitive" *characteristics* such as the reptilian heart structure are termed *ancestral* by biologists; more "advanced" characteristics such as the four-chambered heart are termed *derived*. For convenience, however, I will sometimes use "primitive" and "advanced" for "ancestral" and "derived."

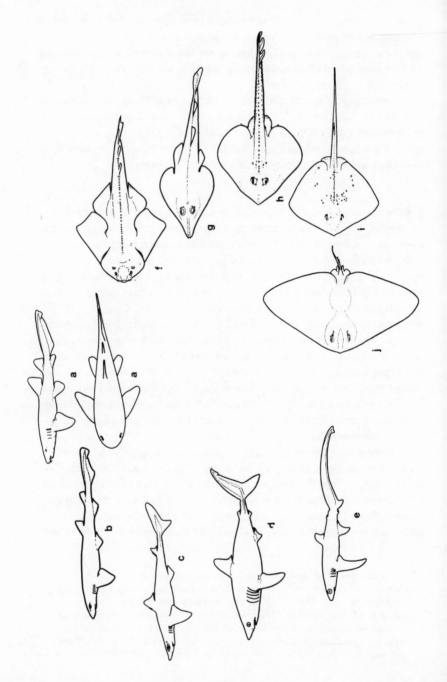

ask how, and from what, they could have evolved. We marvel at the eagle, and wonder that any biologist could seriously suppose its remote ancestors to have been dinosaurlike reptiles.

The problem looks insuperable if we compare an eagle with a lizard, but it doesn't look nearly so difficult if we consider the full spectrum of birds, in all their diversity, and see the species that connect one group of birds to another. The steps from eagle to vulture, from vulture to turkey, and from turkey to the first birdlike fossil *Archaeopteryx* are not all that great; and the step from *Archaeopteryx* to small dinosaurs is quite slight. The pre-Darwinian naturalists had a saying: "Natura non facit saltum"—nature doesn't take leaps. Darwin adopted it as one of the major themes of *The Origin of Species*. Again and again, he emphasized that evolution must proceed by gradual changes of preexisting characteristics, pointing to the intermediate series of species that connect the most disparate of forms. "Look at the family of squirrels. Here we have the finest gradation from animals with their tails only slightly flattened, and from others . . . with the posterior parts of their bodies rather wide and with the skin on their flanks rather full, to the so-called flying squirrels" in which a broad fold of skin between the legs on each side serves as a parachute. Such examples could be given by the thousands. Look at the wasps, and find a series of species in which a structure used to lay eggs is modified more and more perfectly into a sting; at marine

FIGURE 7. An example of two major taxonomic groups, the sharks and the rays, between which the division is arbitrary. The nurse shark (*a*, side and top views) and the angel shark (*f*) have a somewhat flattened body form and live on the sea floor. Whether the angel shark is classified as a shark or a ray is rather arbitrary. The flattened body form, and the joining of the pectoral fins to the head, forming a pair of "wings," become increasingly pronounced in the rays (*g*, guitarfish; *h*, skate; *i*, stingray; *j*, butterfly ray). The sharks illustrated here (*b*, dogfish shark; *c*, scyliorhinid shark; *d*, mackerel shark; *e*, thresher shark) form a series in which the successively more streamlined body and longer fins are adaptations for rapid swimming in open water. The thresher shark and butterfly ray are very different-looking animals, but have developed by successive slight changes from a common ancestor that was probably much like an angel shark or nurse shark. (Dorsal view of nurse shark original. Other figures rearranged from J. S. Nelson, *Fishes of the World*. Copyright © 1976, John Wiley & Sons, Inc. Reprinted by permission of John Wiley & Sons, Inc.)

birds, and find the wings not at all, or slightly, or greatly modified for swimming.

One of the most remarkable revelations of comparative anatomy, in fact, is how seldom truly novel structures are found. We can imagine cherubs and flying horses with wings sprouting from their shoulders; but the wings of vertebrates are always modifications of the front legs. As Darwin's colleague Milne Edwards expressed it, "Nature is prodigal in variety, but niggard in innovation." Take any major group of animals, and the poverty of imagination that must be ascribed to a Creator becomes evident. For example, *all* of the peculiarities of the various modern mammals are simply modifications of the structures possessed by primitive insectivorous mammals such as hedgehogs; and these in turn are modified reptilian features.

If you ask, "What would I have to do to transform a primitive mammal into a bat or a whale?" the answer is, "Nothing very drastic." Bats didn't evolve wings by inventing new structures: the wings are merely elongated fingers, with the same number of joints as in those of a hedgehog, and with an interdigital webbing grown out to the fingertips. The rest of a bat's skeleton is very similar to a shrew's. Whales are an even more striking case. Most whales, such as porpoises, are rather small. Their muscles and a thickened layer of fat give them a streamlined shape. The hind legs are reduced to vestigial pelvic bones. The front legs are flattened into paddles, with five digits (like primitive mammals); but the number of joints per digit is increased. The teeth are partly (in fossils) or entirely (in most modern whales) dedifferentiated, so they all have the same shape; and in modern (but not early fossil) species are increased in number—or else entirely lost, as in the blue whale. The most radical difference from other mammals consists largely of a forward extension of the jawbones out from under the nostrils, which are therefore situated on top of the head. In species such as the blue whale, the skin on the roof of the mouth is cornified like our calluses, and folded into sheets of baleen ("whalebone") that hang down into the mouth. The *only* characteristics that are not mere modifications of primitive mammalian features are the baleen and the dorsal and tail fins, which are rigid folds of skin and fibrous tissue, like our ears.

One of the most amazing aspects of evolution is how easy it is to account for major transformations through rather simple changes in developmental processes. Most of the differences among different kinds of mammals are quite simply accounted for by changes in the relative rates of growth of different parts of the body. Speed up the elongation of fingers to get a bat's wing; slow down the development of teeth or

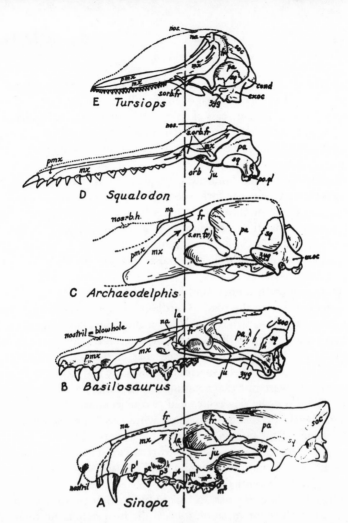

FIGURE 8. The skull of a modern dolphin (*Tursiops,* top), compared to that of fossil whales (*Squalodon, Archaeodelphis, Basilosaurus*) and an Eocene creodont (*Sinopa,* bottom). The creodonts were closely related to the condylarths (see skull of *Phenacodus,* Figure 12), and the whales are believed to have arisen from a condylarth-like stock. *Basilosaurus,* from the Eocene, was thoroughly adapted for aquatic life, but its skull resembles that of the creodont-condylarth group, differing chiefly in the elongation of the premaxillary *(pmx)* and maxillary *(mx)* bones. This trend is carried further in the Oligocene *Archaeodelphis* and *Squalodon,* and even further in the modern whales and dolphins such as *Tursiops.* (From W. K. Gregory, *Evolution Emerging* [New York: Macmillan, 1951]. Courtesy of the American Museum of Natural History.)

legs to reduce or eliminate them in whales; slow down the growth of the lateral toes and increase that of the middle one to get a horse's hoof. In more than 600 million years of animal evolution, there have only been a handful of truly novel features, such as the wings of insects, that seem in our ignorance not to be mere modifications of something that came before.

Darwin held, then (and all subsequent study has borne out his opinion), that a major new form such as a bird does not arise full-blown from a very different ancestor. It arises by the modification of one or a few features that then permit a way of life to which other features become adapted. The earliest fossil birds did not have the hollow bones, enlarged breastbone, and other special adaptations for flight that most modern birds have. Thus, to see how the birds or, for that matter, the apes arose, we try to understand how the characteristics that first led their ancestors down the road to birdhood or apehood were modified from preexisting characteristics. The anatomical knowledge in Darwin's day was enough to show that in most instances, living species run the gamut from primitive to advanced characteristics. Darwin insisted that the differences between species that show the first steps toward new characteristics were just exaggerations of the variations evident within species. To this claim, his critics had two objections.[13] One, raised by Richard Owen, was that the variability within each species was limited. No variations in the fingers of shrews, for example, begin to approach the length of fingers in their supposed relatives, the bats. The solution to this difficulty could not be found until the science of genetics came into being in the twentieth century.*

The other objection was one raised by St. George Mivart. Darwin claims that each slight variation toward the final state is advantageous compared to the preexisting state of the characteristic; but surely there can be no advantage to incipient wings that are insufficient for flight, or to a rudimentary eye that is too imperfect to see with. Natural selection, said Mivart, cannot account for the first few steps toward structures that must be fully formed to be functional.

Darwin had several answers to this objection, and more modern research has both corroborated and modified his views. For one thing, as Darwin realized, not every step of every evolutionary change has to be impelled by natural selection. Stephen Gould and Richard Lewontin of Harvard have recently summarized some of the nonadaptive aspects

---

*I treat this subject in Chapter 7.

of evolution.[14] One such is Julian Huxley's concept of allometric growth, which Darwin recognized, although not by that name. The development of different parts of the body is often correlated, because different developing tissues respond to the same chemical factors. Often one part of the body grows faster than another. The result is that large individuals can have a different shape than small ones. Therefore, when a species evolves large size, some of its features are disproportionately larger or smaller than in its ancestor. In deer, for example, the antlers are relatively larger in large individuals. The extreme was reached by the extinct Irish elk, which was the largest of all deer and had by far the most enormous antlers that any deer has ever had. Incipient antlers, or the horn on a rhinoceros's nose, might not have had any function at all in a very small ancestral species; but as body size evolved, they would automatically become big enough to serve a function, perhaps in fighting, and then would become modified by natural selection. Such features would at first be nonadaptive side effects of other changes and become adaptive in their own right only later.

As developmental biology has matured, it has also come to seem possible that Darwin may have been wrong in his belief that "natura non facit saltum." Some few biologists have always held that new characteristics may indeed evolve in sudden jumps, by "macromutations." The chief protagonist of this idea was Richard Goldschmidt,[15] one of the most eminent developmental geneticists of the 1930s. Goldschmidt pointed to mutations in fruit flies that switch development from one channel into another, transforming antennae, for example, into legs. One of these mutations transforms little peglike structures called halteres into wings. Because the halteres of flies are homologous to the rear pair of wings of other insects, Goldschmidt speculated that this mutation represented, in reverse, the process by which two-winged flies had evolved from four-winged ancestors. He pushed his conclusion to extremes, and theorized that each major taxonomic group had arisen as a macromutation, a "hopeful monster" that in one jump had passed from worm to crustacean, or reptile to bird.

Goldschmidt's ideas were vigorously challenged. Evolutionists such as Ernst Mayr,[16] now of Harvard, pointed out that Darwin's idea of gradual change was attested by the spectrum of intermediate species that connect one group to another; that any mutation with so drastic an effect would be an inviable cripple, as the four-winged mutant flies indeed are; and that genetic studies showed the differences between species to be formed from the same slight genetic variations that are found within species.

Was Goldschmidt wrong? It is clear now that his extreme position was; but some evolutionists have begun to argue that there may have been a germ of truth in his position.[17] We now know that slight biochemical changes can sometimes switch development into very different, yet viable and harmonious, pathways. In salamanders, for instance, a simple change in hormone balance prevents the larva from developing adult characteristics, so that juvenile features are retained throughout life. Perhaps, as Stephen Jay Gould believes, slight genetic changes in biochemistry have sometimes altered developmental rates so as to abruptly change the features of organisms and set them off on a new way of life. The evidence for such a hypothesis is not yet sufficient for us to judge how right he may be.

Whatever the merits of these several theories about the origins of new features, there is little doubt that Darwin's own explanation must still hold first place. New organs usually arise from preexisting structures that acquire a new function. That is, function changes; only then, if at all, does a change in structure follow. The water ouzel, a thrushlike bird that walks about under the rushing waters of mountain streams, shows no particular adaptations for aquatic life. It has evolved the behavior of living in water, and could now probably profit from structural modifications that could adapt it further to an aquatic style of life. Many other species of aquatic birds, however, which must have also started out without adaptations for swimming, display a full spectrum of slight or striking adaptations for life in water. Thus the wings of auks are slightly modified for swimming; the wings of penguins are modified entirely into flippers. Wasps and bees did not develop a sting *de novo* in order to protect themselves. They use a modified egg-laying tool that is adapted in their more primitive relatives to insert eggs into plants. The result is that only female wasps and bees can sting, and the males are defenseless. "New" characteristics are already well enough developed to serve a new function because they had already been elaborated to serve a different function.

The enormous masses of data that generations of biologists before Darwin had acquired—on anatomy, embryology, and geographical distributions—made little sense, and had no coherent theory to explain them, until the hypothesis of evolution organized them into a simple framework. With evolution, such diverse facts as the modification of the same bones for different functions in whales and bats, the useless, rudimentary legs of the boa constrictor, the presence of woodpeckers in the plains of Argentina and their absence from the forests of the Galápagos Islands, and the peculiar geographical distribution of lungfishes all fit into a common explanatory framework, just as the hypothesis of continental drift explained in one sweep the complementary shapes of Africa and South America, the profusion of volcanoes around the shores of the Pacific, the marks of glaciers in eastern Argentina, and the fossil reptiles in Antarctica.

Both theories, evolution and continental drift, at first met strong resistance from scientists, and then quickly won them over, because the theory brought together diverse phenomena under a single, simple explanation. Neither the hypothesis of continental drift nor that of evolution was *proved* true before it won acceptance. In fact, no scientific hypothesis is ever proved true. Rather, it is tested continually with more and more data, and as the data continue to conform with the idea, and fail to disprove it, the community of scientists comes to have more and more confidence in its validity. As geologists obtained more and more data from the rocks of the Atlantic seafloor, all of which conformed with the hypothesis of continental drift, they came to regard it as less of a hypothesis and more of a fact; for what is a fact but a thoroughly confirmed hypothesis?

Similarly, within thirty years of the publication of *The Origin of Species*, there was hardly a biologist alive who didn't regard evolution as a fact—for data from all fields of biology poured in, all thoroughly consistent with the idea that species are descended from common ancestors. Entirely new fields of biology—behavior and biochemistry, for example—came into existence, and all their data likewise made sense in terms of evolution. No one observation or experiment has "proven" evolution, any more than one experiment has proven that the characteristics of living organisms are due to the molecules of which they are made. Both ideas—that life is chemistry, and that life has history—have become scientific "facts" because with them biology makes sense; without them it is chaos.

FOUR

THE FOSSIL RECORD

The business of proving evolution has reached a stage when it is futile for biologists to work merely to discover more and more evidence of evolution. Those who choose to believe that God created every biological species separately in the state we observe them but made them in a way calculated to lead us to the conclusion that they are the products of an evolutionary development are obviously not open to argument. All that can be said is that their belief is an implicit blasphemy, for it imputes to God appalling deviousness.
—THEODOSIUS DOBZHANSKY, 1962*

"If it be said that these species be lost out of the world: that is a supposition which philosophers hitherto have been unwilling to admit, esteeming the destruction of any one species a dismembring the uni-

---

* Theodosius Dobzhansky, perhaps more than any other person, elucidated the genetic basis of evolutionary change.

verse and rendring it imperfect, whereas they think the divine providence is especially concerned to preserve and secure all the works of
creation."

That was the reaction of John Ray, author of *The Wisdom of God
Manifested in the Works of the Creation,* when in 1713 he was faced with
the reality of fossils that represented no known living species. Thomas
Jefferson's reaction to mammoth bones was the same in 1785: "Such is
the economy of nature, that no instance can be produced of her having
permitted any one race of animals to become extinct." Jefferson, an
ardent amateur paleontologist, believed that mammoths must still exist
in the unexplored regions of the far north, and held a similar opinion
of the giant ground sloth that he described from bones exhumed in
Virginia. The philosophy of centuries past could not admit extinction:
if God were perfect and all-good, his creation must have been perfect.
How, then, could he allow it to disintegrate? How could he let the great
chain of being be broken?[1]

By the early nineteenth century, the evidence of extinct animals was
undeniable. How could they be explained? The Biblical flood, of course,
was the obvious possibility; but geology showed that there had been
many periods of extinction. Not one catastrophe, but a multitude, must
be invoked. However, against the idea of successive catastrophes, a new
and powerful idea arose and triumphed. James Hutton in 1788 argued
powerfully for the principle of uniformitarianism: the idea that the
processes we observe today have always operated. Mountains and hills
are eroded into sediments; sediments are compacted into rocks; rocks are
uplifted as volcanoes arise and mountains are formed. "Every material
being exists in motion . . . rest exists not anywhere," said Hutton, and
argued that the past conditions of the earth can be inferred from the
present. But these processes operate so slowly that the earth's formations
must have taken place not in six thousand years, as a literal interpretation of the Bible would have it, but over aeons: "we find no vestige of
a beginning—no prospect of an end."

Hutton's ideas were violently rejected, and almost passed into oblivion for thirty years. In the 1830s, however, they found a powerful advocate in Charles Lyell, whose *Principles of Geology* became one of the
most respected scientific works of the age, and the strongest single
influence on Darwin's ideas. Lyell brought together masses of minutiae
in his argument for uniformitarianism: the fossilized marks of raindrops,
for example, showed "that the atmosphere of one of the remotest periods
in geology corresponded in density with that now investing the globe."
He convinced the scientific world that, as one review of his book put

it, "the concession of an unlimited period for the working of the existing powers of nature has permitted us to dispense with the comets, deluges and other prodigies which were once brought forward, *ad libitum,* to solve every difficulty in the path of the speculating geologist."

Geology, then, showed that the world was immensely old—old enough to permit the slow transformation of species that Darwin would argue had given rise to the diversity of life. It gave evidence that far more species had perished from the earth than are alive today, and that with the passage of time new species came into being. It also established that there was an orderly sequence of fossils: that the history of the earth was marked by changes in fauna that were correlated from one place to another, and could be used to establish geological eras and periods. In time, the present geological table was established—not by evolutionists, but by geologists who believed in successive creations.[2]

Yet until rather recently, the geological time scale was a relative, rather than absolute, chronology. The early estimates of absolute age, based on rates of sedimentation and other geological processes, have in fact turned out to be underestimates. A technique for absolute dating, the use of radioactive isotopes, did not emerge until the early part of this century.[3] This technique is based on the discovery by nuclear physicists that certain atoms, or "parent nuclides," become spontaneously transformed into stable "daughter nuclides" by the loss or addition of protons, neutrons, or electrons. The most remarkable aspect of this process is that it occurs at an absolutely constant rate, independent of temperature or other environmental conditions. Thus out of every 100 potassium-40 nuclides, 50 will decay to argon-40 in 1,300 million years; in the next 1,300 million years, 25 of the remaining 50 will decay to argon-40, and so on. By measuring the relative amounts of parent and daughter nuclides in a volcanic rock, its age can be easily calculated. Because different radioactive isotopes decay at different rates, possible inaccuracies in radioactive dating can be cross-checked by using several different isotopes.

With this method, geologists have obtained the dates for the geological ages as shown in the Geological Time Scale (page 73). They have traced the successive emergence of the Hawaiian Islands, from the oldest, Kauai, to the youngest, Hawaii (which is a mere 100,000 years or so old). They have dated ancient earth rocks, moon rocks, and meteorites, and found a consistent age for the solar system. They have corroborated their dates in some most surprising ways. Astronomers have postulated that because of tidal friction, the rate of the earth's rotation has slowed down at the rate of two seconds every 100,000 years, so that a Paleozoic day should have been about twenty-one hours long. Corals lay down a layer in their skeleton every day, as well as layers that mark the passage of years. John Wells of Cornell University reasoned that Devonian corals, if they really lived 380 million years ago, should have about 400 daily layers in their skeleton for each annual layer—since there must have been about 400 days per year in the Devonian. And so they had; the estimate of the age of the Devonian, deduced from coral skeletons, corresponded perfectly with the estimate from radioactive dating.[4]

Despite such striking corroborations of these dating methods, the "scientific" creationists continue to claim that the earth and the universe came into being a few thousand years ago. They flatly deny that radioactive dating methods are trustworthy, citing possible inaccuracies, even though every geology library has dozens of feet of shelf space devoted to the intricacies of the dating method, the ways in which the level of inaccuracy can be determined and dates can be cross-checked.

The other claim of the creationists is that we have no way of being sure that the rate of radioactive decay has always been constant, even if it seems to us now to be immune to any outside influences. This is the most absurd objection of all. The very same processes of atomic change that result in radioactive decay are those that enable us to build atomic bombs and nuclear reactors. The physics of these processes is very well understood—perhaps too well. Physicists have found that these processes are responsible for the "evolution" of the elements that occurred when the "big bang" formed the universe about 14 billion years ago. The fusion of hydrogen atoms into helium atoms, which is occurring right now in the sun, is the same kind of process that gave rise to all the rest of the elements during the big bang. Thus the ratios of various elements and isotopes in the stars are one line of evidence about the age of the universe.[5] Independent evidence comes from the trajectories of the stars, and the velocities and distances of the receding galaxies.

It is, of course, conceivable that atoms have not always decayed at the same rates that they do now. It is also possible, by the same argument, that hydrogen and oxygen haven't always reacted to form water, or that energy hasn't always been conserved. If scientists reject the principle of uniformitarianism, they can no longer do science. It is absolutely essential to assume that chemical reactions occurred a thousand years ago the same way they do today, and that atomic decay has always followed the same principles that it now does—unless we have some good reason to suppose otherwise. Physicists have yet to find anything that can alter the rate of radioactive decay, so we have no reason to think that it has ever been different.

Given this, what are the major features of earth history? At the present time, we believe that the universe came into existence about 14 billion years ago.[6] An explosion of inconceivably dense matter "created" elements, and sent stars hurtling in all directions into space on trajectories that they are still following, in an ever-expanding universe. Our solar system, including the earth, originated about 4.6 billion years ago. The atmosphere of the early earth was very different from the present one. There was no free oxygen (O_2) or ozone (O_3), so the earth received an enormous input of ultraviolet light that would be lethal to modern life. Ancient Precambrian* rocks are rich in unoxidized iron compounds that could not have been formed in the presence of oxygen.[7]

Among Precambrian rocks about 3.7 billion years old are some that contain peculiar layers of oxidized iron that almost certainly evince the presence of life, because they resemble deposits like those that are formed today by iron-using bacteria. The oldest unquestionable fossils, however, are bacterialike forms in South African strata 3.2 to 3.4 billion years old. By 2 billion years ago, similar fossils, including blue-green algae (now called cyanobacteria), are abundant. Doubtless they produced abundant oxygen by photosynthesis. From this time on, the rocks give evidence of an oxygen-rich atmosphere.

The bacteria and blue-green algae have a much more primitive cell structure than green algae and other plants and animals, which are called eukaryotes. The first green algae do not appear in the fossil record until about 1.6 billion years ago; the "invention" of the modern cell was apparently so unlikely that more than a billion years passed before it made its appearance. By about 700 million years ago there was clearly a great diversity of animal life. The fauna from these rocks is very sparse, but it contains worms, coral-like animals, and an occasional trilobite.

Animal fossils do not appear in profusion, however, until the beginning of the Cambrian period, 580 to 600 million years ago, and within the next 50 million years or so all the animal phyla that have fossilizable skeletons appear in the geological record. At first glance, it seems as if all the major groups of animals arose in a very short time, but this is clearly an illusion; 700-million-year-old Precambrian rocks have a rather diverse fauna, and the very fine-grained Cambrian shales of British Columbia show that there was an enormous diversity of animals that

---

* The accompanying table (page 73) shows the geological time scale, with the names of the eras, periods, and epochs.

Millions of Years Ago	Era	Period	Epoch	Major Events
0.01 —	Cenozoic	Quaternary	Recent	Development of human agriculture and civilizations.
2 —			Pleistocene	Successive ice ages. Extinction of mammoths and other species. *Homo sapiens* widespread.
12 —		Tertiary	Pliocene	Human evolution. Diversification of antelopes, rats, etc. "Modern" elephants, horses.
25 —			Miocene	Diverse families of mammals arise, others die out.
36 —			Oligocene	Many modern mammal families, e.g. horses and rhinoceroses, dogs and cats, become distinct.
58 —			Eocene	Most modern orders of mammals distinguishable, but modern mammal families not distinct.
63 —			Paleocene	Great diversification of primitive mammals; modern orders not distinguishable.
135 —	Mesozoic	Cretaceous		Explosive diversification of flowering plants, modern insects. Major extinctions, including last dinosaurs. Continents fairly well separated.
181 —		Jurassic		Dinosaurs diverse. First birds and first traces of flowering plants.
230 —		Triassic		Continents begin to separate: Diversification of dinosaurs and other reptiles begins. Pine-like plants (gymnosperms) dominant. Some primitive mammals.
280 —	Paleozoic	Permian		Diverse reptiles, including mammal-like reptiles. Major extinction of many invertebrates and vertebrates at end of period. Continents aggregated into one.
345 —		Carboniferous		Diversification of amphibians, fern-like plants. First reptiles.
405 —		Devonian		Diversification of fishes. First amphibians and insects.
425 —		Silurian		First terrestrial plants and arthropods.
500 —		Ordovician		First jawless vertebrates (fishes).
600 —		Cambrian		Appearance and rapid diversification of most animal phyla.
	Precambrian			Origin and diversification of algae and other one-celled organisms. Diverse animals toward end of era.

lacked skeletons. Very possibly this "rapid" diversification of animals in the Cambrian was due to the rapid evolution of hard parts by groups that had evolved long before.[8]

Various groups of invertebrate animals arose and became extinct in subsequent geological ages. The squidlike animals known as ammonoids, for example, underwent several periods of diversification from the Devonian until the end of the Cretaceous, about 65 million years ago, when the last of many hundreds of species became extinct. The first vertebrates, fishlike creatures without jaws, arose about 500 million years ago and were followed by successive radiations of various groups of fishes, including the crossopterygians. These fishes included species with skulls and other skeletal characteristics almost identical to those of the first amphibians, which arose from the crossopterygians in the Devonian. The last fossil crossopterygian is in 70-million-year-old rocks, but in 1937 a living crossopterygian, the coelacanth, was discovered in the Indian Ocean. Thus a group can persist for 70 million years and not be found in the fossil record—which shows how incomplete the record is.

By the Carboniferous period, diverse amphibians were crawling through dense forests of primitive fernlike plants. Not the slightest sign of flowering plants—not a leaf or a pollen grain—appears until well into the Jurassic, almost 200 million years later. The first few reptiles evolved from amphibians in the Carboniferous, and diversified in the Permian, when they included many species known as therapsids, or mammal-like reptiles. Whether their legs are splayed out to the sides or held straight under the body; whether their teeth are all similar or their shapes are differentiated along the jaw; whether their quadrate bone supports the lower jaw or is reduced to a small bone in the middle ear; whether their lower jaw consists of several bones as in reptiles or a single bone as in mammals: the therapsids show a complete gradation from reptiles to mammals in every skeletal detail that paleontologists use to define the class Mammalia.[9]

However, the mammals themselves didn't diversify into their modern groups until over 150 million more years had passed. First came the great extinction at the end of the Permian, during which it is estimated that more than half of the kinds of animals in the world perished; then the great Age of Reptiles, the Mesozoic Era, in which dinosaurs large and small flourished. By then, the continents, which had formed the great landmass of Pangaea, were starting to separate. The last of the dinosaurs perished in the second great episode of extinction in earth history, at the end of the Cretaceous, about 65 million years ago. Before they went, however, they bequeathed to us the birds. The first known bird, *Archaeopteryx* in the mid-Jurassic, resembles certain of the very small dinosaurs in almost every feature except its feathers.[10] In fact, several museum specimens of *Archaeopteryx* were overlooked for years because their skeletons were so similar to those of other Jurassic reptiles.

During the Age of Reptiles, mammals existed, but their fossils are few. Perhaps their diversification was prevented by the proliferation of reptiles, which filled so many ecological niches. In any case, various primitive small mammals are known mostly from scattered teeth throughout the Mesozoic, and not until the Paleocene epoch, 63 million years ago, did they begin to diversify into the modern orders of mammals. The radiation of mammals at this time was very rapid. The Paleocene rocks contain a great diversity of primitive forms that vary in subtle

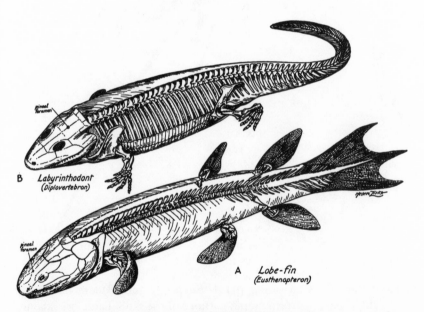

FIGURE 9. An Upper Devonian lobe-finned fish, *Eusthenopteron*, and a Carboniferous amphibian, *Diplovertebron*. Although there are a considerable number of differences between them, the skulls are very similar in structure, and many of the bones in the paired fins of the lobe-fin correspond with those in the legs of the amphibian. (From W. K. Gregory, *Evolution Emerging* [New York: Macmillan, 1951]. Courtesy of the American Museum of Natural History.)

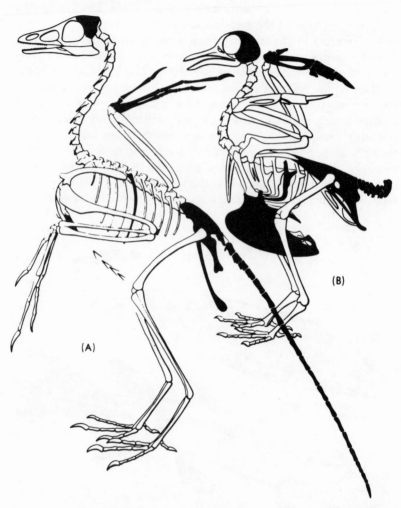

(A)

(B)

FIGURE 10. The skeletons of (A) *Archaeopteryx* and (B) a modern bird, the pigeon. Corresponding portions of the skull, hand, breastbone, rib, pelvis, and tail are in black. Note that in the pigeon, the bones of the hand are fused, as are those of the back vertebrae, and the breastbone is greatly enlarged. The ankle bones of *Archaeopteryx* are separate, unlike the pigeon. Compare *Archaeopteryx* to the small dinosaur *Coelophysis* in Figure 11. (From E. H. Colbert, *Evolution of the Vertebrates*, 3rd ed. Copyright © 1980, John Wiley & Sons, Inc. Reprinted by permission of John Wiley & Sons, Inc.)

ways. Many are rather like shrews and hedgehogs, the insectivores that are considered the most primitive of the mammals with a placenta. Some of the primitive Paleocene mammals slightly resemble primates; others resemble weasel-like carnivores.

Most of the modern orders of mammals are represented by less specialized species as we go back in time, until when we reach the Paleocene, they become so unspecialized that it's harder and harder to distinguish one from another. The condylarths, for example, appear to be ancestral to various groups of hoofed animals; but the condylarths are similar to the creodonts, which appear to be primitive carnivores; and many of the creodonts could equally well be classified as insectivores.[11]

Three features of this story are critical to our present purposes. First, there is immense regularity in the fossil record. Mammoths, dinosaurs, and trilobites aren't mixed together at random. From the beginning of the Cambrian, hundreds of millions of years pass before the first amphibians appear; then another hundred million or so years until the first reptiles, and another hundred million until the first birds. Without any reference at all to the fossil record, taxonomists have claimed, using the principles by which they construct phylogenetic trees, that modern mammals are descended from primitive shrewlike insectivores, primitive mammals from reptiles, reptiles from amphibians. These judgments come entirely from anatomical studies of living species. We predict,

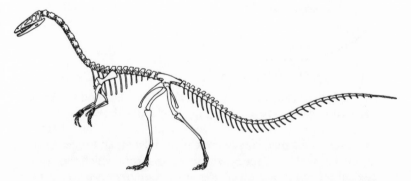

FIGURE 11. One of the Triassic theropod dinosaurs, *Coelophysis*. This was a small dinosaur, about eight feet long, but even smaller species existed. In most major respects the skeleton is very similar to that of *Archaeopteryx* (Figure 10). (From E. H. Colbert, *Evolution of the Vertebrates*, 3rd ed. Copyright © 1980, John Wiley & Sons, Inc. Reprinted by permission of John Wiley & Sons, Inc.)

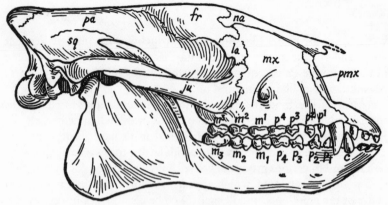

Phenacodus

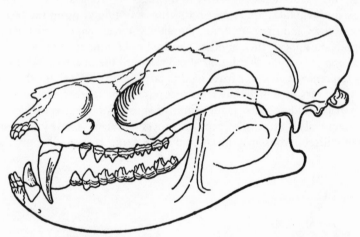

Deltatherium

FIGURE 12. A Paleocene creodont, *Deltatheridium* (bottom) and an Eocene condylarth, *Phenacodus* (top). Although the creodonts were somewhat modified for a carnivorous diet and the condylarths for an herbivorous diet, they were closely related, and together constituted a stock of unspecialized mammals from which the modern hoofed mammals and carnivores arose. Compare *Phenacodus* to its close relative *Hyracotherium* (Figure 16), the ancestor of the horses. The skull of *Deltatheridium* (like that of another creodont, *Sinopa*, which is illustrated in Figure 8) resembles that of some of the more primitive modern carnivores (see Figure 13). (From W. K. Gregory, *Evolution Emerging* [New York: Macmillan, 1951]. Courtesy of the American Museum of Natural History.)

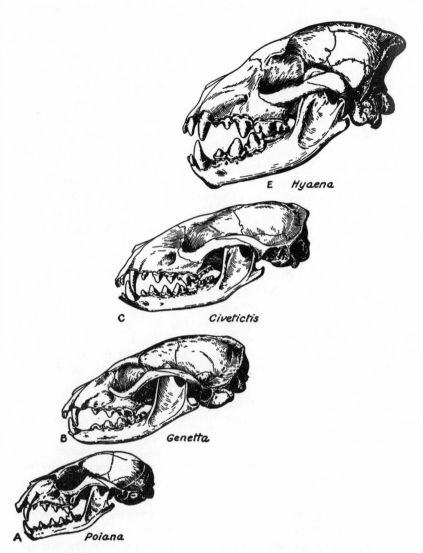

FIGURE 13. Skulls of some modern carnivores. The three lower forms are members of the family that includes civets and mongooses; note their resemblance to the Paleocene creodont *Deltatheridium* in Figure 12. The hyaena, at top, is modified from the civets chiefly by the more robust skull and jaw and the larger teeth, which are adaptations for feeding on bones. These are all living species which are derived from common ancestors rather than from one another. (From W. K. Gregory, *Evolution Emerging* [New York: Macmillan, 1951]. Courtesy of the American Museum of Natural History.)

then, that amphibians, reptiles, primitive mammals, and modern mammals should appear in sequence in the fossil record, and they do. It is impossible, if evolution is true, that any mammal fossils should ever be found in Devonian rocks, and indeed there are no such fossils.

Second, the methods of reconstructing phylogenetic trees from living species often enable us to determine which characteristics are ancestral and which are derived. For example, comparative anatomy tells us, independently of the fossil record, that the pair of occipital condyles in mammals (the pair of bumps that join the skull to the vertebrae) is advanced, compared to the single condyle in reptiles. We predict, then, that fossil reptiles and the earliest mammal-like fossils should have a single occipital condyle, and they do.[12] That is, the farther back we go in the fossil record, the more we find organisms with the ancestral characteristics we expect.

Third, as we pass from the remotest periods of geological time toward the present, the fossils become more and more modern. Certainly some groups such as blue-green algae and horseshoe crabs have persisted since early geological time; but most groups of animals and plants have arisen, flourished, and died out. The most ancient fossils are the strangest to us; and as we approach the present, they get more and more familiar. Jurassic mammals, if they came alive today, would hardly look like mammals to us; by the Cretaceous, we get rather modern-looking opossums; by the Eocene, armadillos; by the Pliocene, modern-looking horses and rhinoceroses. Regularities of this kind accord with evolution, not creation.

The fossil record is not, of course, a book that we can open at will to look up dates and historical figures. It is the accumulation, by hard work, of the fragments of early life that happen to have been preserved and happen to have been found.[13] No locality on earth has a complete series of geological strata from the Precambrian to the present; there are gaps of hundreds of millions of years between, say, Devonian deposits and Cretaceous deposits that may lie immediately above them. And even within a Cretaceous bed, sediments were not laid down continually: successive layers are usually separated by thousands or hundreds of thousands of years. Thus the fossil record is a source of endless frustration. The museums of the world hold millions of fossils, but they are from rich beds found here and there, a scattering of fragments from the vast expanse of time and space.

Moreover, we know that the recovery of any organism from the past depends on a concatenation of improbable events: the organism must have hard parts that resist decay; it must be buried in sediments that happen to become solidified into rock; the rock must escape erosion and metamorphosis for eons; and it must be exposed in places where geologists happen to find it. Many fossil species that must have been abundant are known from only a single tooth or a few skeletons. Recently, for example, the *New York Times* reported that Harvard paleontologist Farish Jenkins has found the jaw of a shrewlike mammal from upper Triassic rocks in Arizona. Although the fossils of a few small mammals of this age had been found in Europe, this is the earliest mammal yet known from North America. How many more are waiting to be found?

Poor as the fossil record is, however, it tells us that there is an orderly history of life. Different groups originated at different times, not all at once. It is inconceivable that flowering plants or horses could have existed for hundreds of millions of years without leaving a trace, and then left a rich record thereafter. The rocks tell us, also, that extinction is the fate of almost all species. Moreover, the rate of extinction doesn't slow down as time goes on; recently evolved species have no longer a tenure on earth than ancient ones.[14] This implies that mammals, for example, are no better adapted to withstand the changes that time inflicts upon them than clams or reptiles were. In fact, an average genus of mammal lasts in the fossil record for only a tenth the time that an average genus of clam does. This implies that species are adapted to immediate environments, and have not become better during evolution at withstanding unforeseen changes in the environment.

The fossils also show that the progenitor of a modern group usually has only one or two of the key features that typify the group in its later history. The first horse, the "eohippus" (properly called *Hyracotherium*), had a peculiarity of an ankle bone that is common to horses, rhinoceroses, and tapirs that make up the order Perissodactyla; but in almost no other respect did it resemble a modern horse.[15] Sometimes we can use such evidence, with caution, to distinguish ancestral from derived characteristics. For example, we can safely say that the single jawbone of mammals evolved from the multiboned jaw of reptiles, because all fossil reptiles, including most therapsids, had several bones in the jaw. This evidence must be used with caution, though, because a fossil may be a specialized species, with its own peculiarities, that never gave rise to modern species. The peculiarities of dinosaurs are not ancestral to mammal characteristics, because other reptiles, not dinosaurs, gave rise to mammals.

The fossil record also shows that rates of evolution can vary enormously. Bjorn Kurtén has shown that in bears and many other mammals, characteristics such as teeth can fluctuate rapidly in size, increasing and decreasing appreciably within a few hundred thousand years.[16] (A geologist's concept of time is breathtaking; geologists talk of a "few million years" as casually as we speak of the day before yesterday.) In other cases, the characteristics of a group persist virtually unchanged for many millions of years. The opossum, for example, is hardly distinguishable from a 70-million-year-old Cretaceous species. The blue-green algae have existed since the Precambrian, in which fossils indistinguishable from modern forms are found.

As far as we can tell, species possess no intrinsic drive to evolve, no impetus toward progress. If they are sufficiently adapted for an environment that persists through long stretches of time, their adaptations may also persist without change. A very common pattern, in fact, is for a group to evolve very rapidly at first, and then to level off after their new adaptations have been more or less stabilized in a final form. The lungfishes, for example, evolved rapidly in the Devonian, and by the beginning of the Permian they reached an adaptive "plateau" that they have stayed on ever since.[17] This pattern suggests that in order to trace the gradual evolution of a major new group, it is necessary to find fossils from that relatively brief period in which the new adaptations arise, before they become stabilized.

This principle bears on one of the most striking and potentially embarrassing features of the fossil record. The majority of major groups appear suddenly in the rocks, with virtually no evidence of transition from their ancestors. This is one of the major points of attack by antievolutionists. Almost the entirety of Gish's book *Evolution: The Fossils Say No!* is devoted to these gaps, and the conclusion that if paleontologists can't show gradual evolution, evolution must not have happened. His case sounds devastating. He supports it with quotations from eminent paleontologists ranging from George Gaylord Simpson to Niles Eldredge and Stephen Jay Gould, who, with Steven Stanley, are the chief proponents of a concept they have called "punctuated equilibria."[18]

Eldredge, Gould, and Stanley take their point of departure from Ernst Mayr, who proposed in 1954 that new species must often arise by rapid evolution in very small, localized populations. Mayr arrived at this conclusion not from the fossil record, which he has never studied, but from his studies of birds on islands. Again and again, he found cases in which a species is quite uniform over broad continental areas but is represented by very divergent populations on small nearby islands. He proposed therefore that an abundant, widespread species may evolve only very slowly, and persist virtually unchanged, while small, isolated populations of that species may experience major new evolutionary changes and form new species. Rapid evolution would therefore be coupled with the multiplication of species. When such a newly formed species extends its range, it will overlap with its ancestor.

Whether or not Mayr's hypothesis is right is a matter of some controversy. But if he is right—and it seems very possible that he is— the implications for the fossil record are obvious. The transitional forms that evolve so quickly, and in such a small area, are very unlikely to be picked up in the fossil record. Only when the newly evolved species extends its range will it suddenly appear in the fossil record. Eldredge and Gould have suggested, therefore, that the fossil record should show stasis, or equilibrium, of established species, punctuated occasionally by the appearance of new forms. Hence, the fossil record would be most inadequate exactly where we need it the most—at the origin of major new groups of organisms.

The idea of punctuated equilibria is controversial, among both geneticists and paleontologists.[19] All would agree, however, that rates of evolution can be very rapid. Indeed, laboratory experiments show that a great deal of evolutionary change can occur in twenty or thirty generations—far, far less than the "short" span of a million years or so in which a paleontologist would hardly hope to find transitional forms.[20] It is surprising, in fact, that the evolution of major new forms should ever take anything like a million years. The fastest major changes in the fossil record, such as the evolution of horse teeth, are thousands of times slower than evolution in the laboratory, but on a geological scale they are still very rapid. Evolution is by no means the ponderously slow, stately process that the elementary biology textbooks teach.

Whether or not the fossil record reveals gradual evolution is very much a matter of scale. Continuous geological deposits that last for more than a few hundred thousand years are hard to find. In this interval of time, we would only expect to see species transformed into quite similar species, not into a distinct new family or order. Indeed, a number of fossil sequences do show a gradual change from one species to another in the course of a few hundred thousand years. It is impossible to find an absolutely continuous gradation from an ancestral species to a new family or order—eohippus to the modern horse, for instance—because this requires millions of years, and no geological deposits extend continuously for such a long period. Thus, for short geological intervals we can find continuous evolutionary change, but only to a limited degree. For long intervals we can find examples of great evolutionary change that are roughly gradual over long periods of time, but the record in such cases could not possibly document every minuscule change that may have happened along the way.

For example, the history of horse evolution covers 60 million years. The overall picture is clear: as time passes, horses got larger, lost their side toes, and grew longer teeth. But we don't have fossils every few thousand years, so we cannot tell if this was an absolutely smooth progression. The record suggests that horse evolution is a history of occasional spurts of rapid change, but we cannot be sure.

In contrast, P. G. Williamson[21] has recently described the evolution of new species of snails and clams in some continuous deposits in East Africa. During the late Pliocene and early Pleistocene, a large lake subsided, leaving behind the Turkana Basin, a small isolated lake. The lake's populations of molluscs, which had shown no signs of evolutionary change for hundreds of thousands of years, changed into recognizably similar but easily distinguishable species, going through a gradual change in size and shape. During this process, which took only 5,000 to 50,000 years, the ancestral species remained unchanged in surrounding areas of East Africa. When the lake level rose again, the ancestral species reinvaded the Turkana Basin, and the newly formed species abruptly became extinct.

Williamson interprets this as a case of punctuated equilibrium. Evolution happened during a brief period in which isolated populations diverged from their widespread parents and became new species. Their evolution was gradual, in the sense that they went through intermediate shapes, yet the time span in which it occurred was very rapid. Because most geological deposits do not provide such a detailed record, the sort of gradual evolution that Williamson has described could easily occur during a 50,000-year gap, and it would look as if new species had arisen by sudden jumps.

Some paleontologists do not accept the idea of punctuated equilibria, because many cases of transitions do in fact exist in the fossil record. *Archaeopteryx*, for example, is neither bird nor reptile: it is a reptile with one new key feature—feathers. If, then, we want to see whether a group evolved gradually, we need to ask whether individual features such as feathers evolved gradually. We do not yet have evidence in the case of feathers, but for many other features we have exactly the evidence required. The paleontologist H. K. Erben, for example, has provided striking evidence of the origin of a major, now-extinct group of squidlike animals, the ammonoids, from their ancestors, called bactritids.[22] The ammonoids had a more or less snail-like shell that took on an enormous diversity of forms—tightly coiled, open spiraled, slightly curved, or straight, with all sorts of ridges and complicated sutures. In the early Devonian there is a perfect gradual series from the slightly curved, conical shell of the bactritids (which are not classified as ammonoids) to more and more tightly coiled, elaborate ammonoid shells. The suture pattern, too, shows a sequence of gradually increasing complexity in several different ammonoid groups.

Many such cases exist. Philip Gingerich, at the University of Michigan, has described the gradual increase in the size and form of the teeth of an early Tertiary mammal called *Hyopsodus*. [23] Davida Kellogg, at the University of Maine, has documented cases of gradual size increase in *Pseudocubus*, a marine shelled protozoan, found in continuous deposits that span more than 2 million years. [24] The gradual transition from therapsid reptiles to mammals is so abundantly documented by scores of species in every stage of transition that it is impossible to tell which therapsid species were the actual ancestors of modern mammals. [25]*

The most famous case of evolution, featured in every biology book and every museum exhibit on evolution, is that of the horses. But the story of horse evolution, which was worked out primarily by W. D. Matthew, R. A. Stirton, and G. G. Simpson, is by no means simple. Because its complications are usually ignored by biology textbooks, creationists have claimed that the horse story is no longer valid. However, the main features of the story have in fact stood the test of time, and are worth recounting in some detail. [26]

In the Paleocene of western North America, a diverse group of dog-sized animals known as condylarths existed. The various species show transitions to such groups as the carnivores, the titanotheres, and the rhinoceroses (which at that time included both small running forms and amphibious species). One of the Eocene condylarths was *Phenacodus*, which had five toes with claws that were slightly developed into hooves. The central toe was also slightly enlarged. A close relative of *Phenacodus* was *Hyracotherium*, the Eocene eohippus or "dawn horse." *Hyracotherium* was much like *Phenacodus*, except that it had only four toes on the front foot and three on the rear, slightly longer foot bones, and a slight tendency for the cusps on the molar and premolar teeth to be united into crests. It is important to point out that the differences between *Phenacodus* and *Hyracotherium* are equivalent to those that can

---

* This transition is described in more detail in Chapter 10.

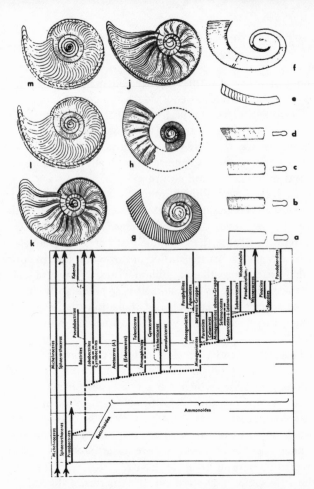

FIGURE 14. Gradual evolution in the ammonoids. Forms *a* through *m* arose at successively later times, as indicated in the diagram of geological strata. In this diagram, solid lines represent the known occurrences of each form. Dotted and dashed lines represent presumed ancestor-descendant relationships. Forms *a* through *e* are bactritids, which may be arbitrarily classified separately or as primitive ammonoids. In this series of genera, the shell became progressively more coiled and developed more elaborate sutures. The forms illustrated are keyed to the geological diagram as follows: a, *Sphaerothoceras;* b, *Protobactrites;* c, *Bactrites;* d, *Lobobactrites;* e, *Cyrtobactrites;* f, *Anetoceras (A.);* g, *A. (Erbenoceras);* h, *Teicherticeras;* j, *Mimagoniatites,* zorgensis-Gruppe; k, *Mimagoniatites,* obesus-Gruppe; l, *Anarcestes;* m, *Werneroceras.* (From H. K. Erben, *Biol. Rev.* 14:641–58 [1966], published by Cambridge University Press.)

86

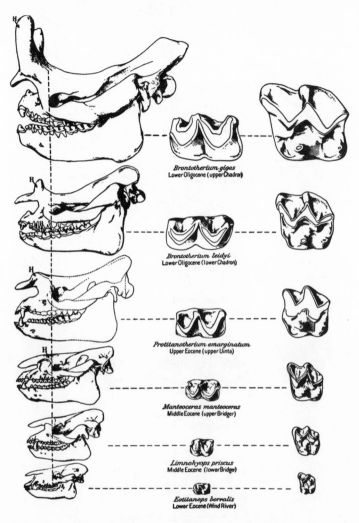

FIGURE 15. One of several series of successive slight changes in the fossil record, illustrating the evolution of the large, rhinoceros-like titanotheres such as *Brontotherium* (top) from smaller, more generalized ancestors (*Eotitanops*, bottom). This series illustrates gradual changes in size, the development of the horns on the nose, and in the form of the crests on the teeth. The horns appear to have developed allometrically, that is, as a consequence of increasing body size. (From W. K. Gregory, *Evolution Emerging* [New York: Macmillan, 1951]. Courtesy of the American Museum of Natural History.)

87

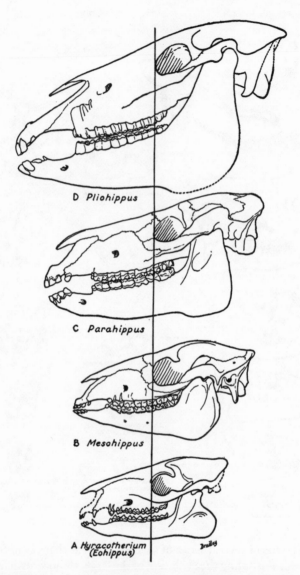

FIGURE 16. Skulls of several of the intermediate forms leading to the modern horse. The vertical line separates the face from the cranial region, showing that in the later species the face became relatively longer. The lower jawbone became deeper to house the elongated teeth situated in deep sockets. *Hyracotherium*, at bottom, is quite similar to the condylarth *Phenacodus*, illustrated in Figure 12. (From W. K. Gregory, *Evolution Emerging* [New York: Macmillan, 1951]. Courtesy of the American Museum of Natural History.)

often be seen within species. The number of toes, for example, varies within many species of living vertebrates.

Throughout the Eocene, for more than 20 million years, most of the characteristics of *Hyracotherium* hardly changed, except for the teeth. The tendency of the cusps to form crests increased continuously, so that the late Eocene form is given a different name, *Epihippus*. Then a slight jump brings us to *Mesohippus*, in the early Oligocene. *Mesohippus* is somewhat larger than *Epihippus*, has a longer face and longer legs, and the first toe (thumb) of the front foot is reduced to a vestigial nubbin. Essentially, *Mesohippus* has three toes per foot, and its side toes are just as large as in *Hyracotherium*. The cusps of the teeth are joined into well developed crests, more suitable for grinding vegetation. Throughout the Oligocene, *Mesohippus* changes gradually into *Miohippus*. It becomes larger, and an extra crest on the teeth that appears first as a variation within the *Mesohippus* population later becomes a typical feature of *Miohippus*.

We cannot be sure why these changes occurred, but we have a fairly good idea. The characteristics of *Hyracotherium* are those of a forest-dwelling animal that browsed on fairly soft foliage and scampered from thicket to thicket. As the Tertiary Period progressed, the climate became drier, and grasslands replaced forests in much of North America. Because of the silica in the leaves, grass is difficult to chew and wears teeth down rapidly. It's likely that the increased ridges of the horses' teeth, and the greater height of the teeth, were adaptations to an increased diet of grass, as they are in certain other groups of mammals such as voles (field mice). In open country, moreover, many mammals escape predators by running long distances, rather than by springing quickly into thickets. The fusion of bones, enlargement of the central toe, and lengthening of the leg that happened in the evolution of the horses provided mechanical advantage for rapid running, as it has in other groups of mammals.

During the next 15 million years, the Miocene, *Miohippus* grades into several distinct lines that diverged from each other. *Archaeohippus* was a dwarfed version of *Miohippus*. *Anchitherium* retained the three-toed condition and the small, simple teeth of *Miohippus*, but there are slight differences in the shape of the tooth crests. *Anchitherium* evolved into a larger form, *Hypohippus*, which became extinct in the early Pliocene. The third line that *Miohippus* gave rise to was *Parahippus* in the early Miocene, a time during which grasslands became more widespread.

In the transition from *Miohippus* to *Parahippus*, several of the smaller crests on the teeth became enlarged, connecting the other crests into a complex series of ridges that were suitable for grinding grass. Moreover, there was a gradual increase in the height of the teeth, so that they could grow continually out of the gums as the tooth surface became worn down. This so-called "hypsodont" condition of the tooth was accompanied by the development of a cementlike substance on the tooth surface, between the ridges. All these changes, as well as an increase in body size and the length of the face, occurred very rapidly, so that *Parahippus* quickly became transformed into *Merychippus*. *Merychippus* had teeth well adapted for grazing grasses, but still had three toes. However, the side toes were large in some specimens of *Merychippus* and very small in others.

By the end of the Miocene there was a great proliferation of species. *Merychippus* split into six different lines that varied in body size and the details of tooth structure. One of these was *Pliohippus*, which resembled one of the species of *Merychippus* in its distinctive pattern of tooth ridges and very small side toes. *Pliohippus* had higher teeth than *Merychippus*, and even smaller side toes. The later species of *Pliohippus* extended these trends even further, and became indistinguishable from the one-toed *Equus*, the modern horse.

The history of horses, then, is very complex, and not at all the steady progress from *Hyracotherium* to the modern horse that is taught in introductory biology books. Evolution didn't follow a straight line; rather, the horses diversified into many different species. Some were adapted for running and grazing on grasses; others were not. As with the small size of *Archaeohippus*, in some species the direction of evolution was even reversed. The path to the modern horse can be traced back in a more or less gradual series over the course of 60 million years. This, however, is only one of many paths of horse evolution, of which all except one ended in extinction.

Considering the entire history of the horse family, it becomes clear that different characteristics such as teeth and toes evolved independently of each other, and that the rate of evolution of each separate characteristic varied greatly. Sometimes tooth height evolved rapidly, as in the *Parahippus-Merychippus* transition, and sometimes slowly, as in *Hyracotherium*. In fact, even the evolution of a single structure such as the tooth is actually an evolution of several different characteristics that may change independently. The height of the tooth and the pattern of cusps and crests are examples. In addition, *Mesohippus* nicely illustrates one of the chief principles of evolution: that the distinctive characteristics of later species originate as variations within the population of the

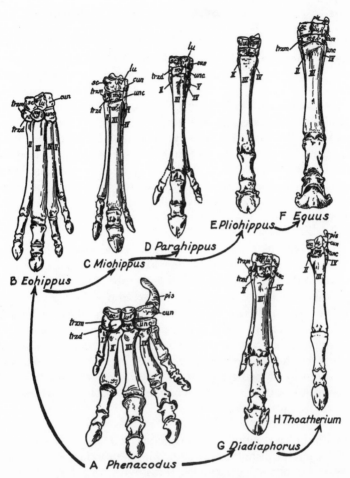

FIGURE 17. Evolution of the front foot in members of the horse family. The side toes of the Eocene condylarth *Phenacodus* were reduced in size, and digit I was absent entirely in its close relative *Hyracotherium* (= Eohippus). The metatarsal bones of *Hyracotherium* were more elongated than in *Phenacodus*. This elongation, and the reduction of the side toes, became progressively accentuated in the North American horse lineage leading to *Equus*, the modern horse. In South America, the "pseudohorses" *Diadiaphorus* and *Thoatherium* underwent a parallel evolutionary change. The foot of *Phenacodus* was more similar to that of *Hyracotherium* than this picture indicates; the digits have been spread out by the artist to show their relationship to the other bones more clearly. (From W. K. Gregory, *Evolution Emerging* [New York: Macmillan, 1951]. Courtesy of the American Museum of Natural History.)

ancestor. The extra tooth crest, which was first a variation within *Meso-hippus,* became one of the distinguishing features of *Miohippus* and its later descendants.

Creationists have claimed that the story of the horse has not held up under modern research. In a long discussion,[27] Duane Gish makes the following points. "George Gaylord Simpson, for example, has declared that several generations of students have been misinformed about the real meaning of the horse." Indeed they were. Simpson, in the book Gish cites, is at pains to point out that there was not one line of horse evolution, but many. "The really striking and characteristic part of the pattern," says Simpson,

> is precisely its repeated and intricately radiating splitting. Its botanical analogue would be more like a bush than like a tree, and even if the tree figure of speech were used, *Equus* [the modern horse] would not correctly represent the tip of the trunk but one of the last bundles of twigs on a side branch from a main branch sharply divergent from the trunk. . . . The whole picture is more complex, but also more instructive, than the orthogenetic [straight line] progression that is still being taught to students as the history of the Equidae. It is a picture of a great group of real animals living their history in nature, not of robots on a one-way road to a predestined end.[28]

Gish also says, without documentation, that "nowhere, for example, are there intermediate forms documenting transition from a non-horse ancestor (supposedly a condylarth) with five toes on each foot, to *Hyracotherium* with four toes on the front foot and three on the rear. [*But* Hyracotherium *is very similar to the condylarth* Phenacodus *in every other respect.*] Neither are there transitional forms between the four-toed *Hyracotherium* and the three-toed *Miohippus* [*but what about* Meso-hippus *with one of the four toes greatly reduced?*], nor between the latter, equipped with browsing teeth, and the three-toed *Merychippus,* equipped with high-crowned grazing teeth [*didn't he read Simpson's book, which describes the gradual change in the teeth of* Parahippus, *the link between* Miohippus *and* Merychippus?*]. Finally, the one-toed grazers, such as *Equus,* appear abruptly with no intermediates showing gradual evolution from the three-toed grazers." But Simpson explicitly describes, in several books, the gradual reduction in the side toes of *Pliohippus,* which connects the three-toed *Merychippus* to the one-toed *Equus.*

Finally, Gish quotes Simpson as saying that *Hyracotherium* is so

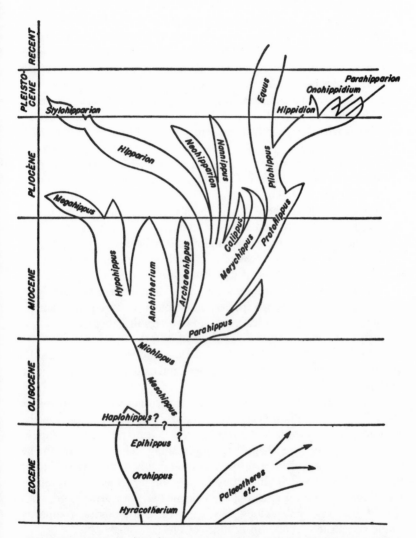

FIGURE 18. A greatly simplified representation of the phylogeny of the horse family, Equidae. The modern horse, *Equus*, can be traced back through *Pliohippus, Merychippus, Parahippus, Miohippus, Mesohippus, Epihippus,* and *Orohippus* to *Hyracotherium,* but this was only one of many lines of evolution. (From G. G. Simpson, *The Major Features of Evolution* [New York: Columbia University Press, 1953].)

primitive that it is no more definitely an equid (horse family) than a rhinocerotid or a tapirid. Gish's interpretation is as follows: "In other words, *Hyracotherium* is not any more like a horse than it is similar to a tapir or a rhinoceros, and thus just as justifiably it could have been chosen as the ancestral rhinoceros or tapir. It seems, then, that the objectivity of those involved in the construction of the phylogenetic tree of the horse was questionable from the very start, and that the 'horse' on which the entire family tree of the horse rests was not a horse at all." But here is what Simpson had to say in *The Major Features of Evolution:*

> For instance, Matthew (1926) pointed out, but later students mostly ignored, the fact that eohippus [*Hyracotherium*] was *not* a horse, that it is about as good an ancestor for *Rhinoceros* as for *Equus*. In effect, there was no family Equidae when eohippus lived. The family and all its distinctive characters developed gradually as time went on. Eohippus is referred to the Equidae because we happen to have more nearly complete lines back to it from later members of this family than from other families. There is no particular time at which the Equidae became a family rather than a genus or a species; the whole process is gradual and we assign the categorical rank after the result is before us.[29]

The point, of course, is that *Hyracotherium* and its relatives were the ancestors of both horses and rhinoceroses, and can't be assigned to either of these groups.

The history of the horses is not perfectly documented, but it is as good as any 60-million-year history is likely to be. The horses, the ammonoids, the mammal-like reptiles, the marine protozoans, and, as we will see in the next chapter, the line leading to humans, give the lie to the creationists' claim that fossils fail to document evolution. Except to one whose reason is blinded by unquestioning adherence to fundamentalist doctrines of creation, the evidence of the fossil record, with that of anatomy, embryology, biochemistry, and genetics, compels a single conclusion: evolution is a fact of life.

But this hardly addresses the knottiest problem of all—the origin of life. Of all the claims of evolutionists, this is the most forbidding. A devoutly religious person may well believe that the Creator instilled into life the ability to vary and to change in response to the environment, and so enabled life to diversify without divine intervention. He or she may, however, balk at the notion that life itself could have developed through natural causes from inanimate matter. If species can evolve and diversify, and if life itself can evolve, and if the structure of the universe can be explained by physicists, where can we find evidence of the Creator's handiwork, except perhaps in the very origin of matter and energy? The implications are so daunting that Darwin himself was reluctant to commit his beliefs to paper. In *The Origin of Species* he limited himself to saying that "probably all the organic beings which have ever lived on this earth, have descended from one primordial form, into which life was first breathed"—a phrase which is certainly open to theological interpretation.

If the word "life" conjures up only images of oak trees, bumblebees, and humans, it certainly seems inconceivable that life should have arisen by the aggregation of primordial molecules. To a biologist, though, "life" encompasses an entire spectrum ranging from such complicated organisms to bacteria and viruses. The simplest forms of life, the viruses and their relatives known as phages, are little more than macromolecules. The phage called $\Phi X174$, for example, consists only of a length of DNA (deoxyribonucleic acid, the genetic material) 5,375 units (nucleotides) long, that contains the information for producing only nine kinds of proteins. Such things are considered to be living because of two essential characteristics: the ability to obtain energy from the environment, and the ability to replicate themselves and produce new copies that are sometimes slightly altered. They have heredity, the essence of life.

We will almost certainly never have direct fossil evidence that living molecular structures evolved from nonliving precursors. Such molecules surely could not have been preserved without degradation. But a combination of geochemical evidence and laboratory experiment shows that such evolution is not only plausible but almost undeniable.

While the primeval atmosphere of the earth lacked oxygen, it must have had methane, ammonia, water vapor, and other elementary gases such as carbon monoxide. Because there was no ozone shield, ultraviolet light provided an intense source of energy. When these conditions are duplicated in the laboratory, as Stanley Miller and Harold Urey did in 1951, and as many other chemists have done since then, an enormous variety of organic molecules are formed spontaneously:[30] sugars, the amino acids that are the building blocks of proteins, and the nucleotide bases that are the building blocks of DNA. Moreover, these amino acids spontaneously assemble themselves into short proteins, which aggregate into spherical polymers that almost look like cells, and split into smaller spheres when they grow too large.

All living things require genetic information in the form of DNA or RNA (ribonucleic acid). They also require proteins known as enzymes, which are encoded in the genetic information and in turn assemble free nucleotides into new copies of DNA or RNA. So far such a complete nucleic acid–plus–enzyme system has not been synthesized in the laboratory, but Manfred Eigen and his co-workers have come very close.[31] They have found that the enzyme that replicates the RNA genetic information of a phage called Q_β can string nucleotides together into short strands of RNA even in the absence of preexisting RNA. Many short sequences of RNA are formed in this way, some of which replicate faster than others, and replace the slower-replicating sequences: that is, they evolve by natural selection. Moreover, different sequences, or "species," of RNA replicate fastest in different chemical environments. Thus, evolving genes can arise in the absence of preexisting life. Leslie Orgel has found, moreover, that even without a replicating enzyme short sequences of RNA can replicate themselves. The next step in the synthesis of life will be to develop RNA sequences that can produce their own replicating enzymes. This hasn't been done yet, but biochemists have made such great progress in this direction that it's likely that simple forms of life will be synthesized in the laboratory within the next ten years.

It is important to realize that although human intelligence is guiding such experiments, chemists are not making RNA molecules by carefully stringing together nucleotides with sophisticated chemical techniques. They are simply providing in the laboratory the chemical and environmental conditions that are believed to have existed naturally billions of years ago.

The genetic code of DNA and RNA is identical in all species from viruses to mammals. Thus all living things share fundamental biochemical characteristics which indicate that they have all evolved from a single form of life. We begin to understand, from laboratory experiments, how that earliest form of life, perhaps a self-replicating nucleic acid, could have arisen spontaneously from simple chemical compounds. Geology shows that life evolved more than 3 billion years ago and has diversified ever since. Many later forms of life are much more complex than the early single-celled forms, but life as a whole shows no consistent directional trends. "Evolutionary progress" is a human construct that we inherit from the Victorian age of optimism. It is not inherent in the process of evolution. Because evolution can occur rapidly at times, and because the fossil record is incomplete, there are many gaps between ancestors and their presumed descendants; but even so, the rocks reveal numerous instances of gradual evolutionary change and show that groups of animals that are quite distinct in the modern world become more and more indistinct as we pass back in time. Thus, together with the evidence of anatomy, embryology, biochemistry, and the geographic distribution of species, the fossils reveal a history of descent with modification.

The history of the earth is a story of billions of years of change: drifting continents, changing climates, uplift and erosion of whole mountain ranges. Millions of new and wonderful species have arisen and diversified, and millions—more than 99 percent of those that have ever lived—have disappeared. Five hundred million years ago the fishes first appeared; 65 million years ago the last of the dinosaurs expired; a million years ago Massachusetts was buried under ice and woolly mammoths walked through Virginia; and a mere 12,000 years ago human agriculture started to change the face of the earth.

FIVE

HUMAN
EVOLUTION

I f we reject the natural explanation of hereditary descent from a common ancestry, we can only suppose that the Deity, in creating man, took the most scrupulous pains to make him in the image of the ape. This, I say, is a matter of undeniable fact—supposing the creation theory true—and as a matter of fact, therefore, it calls for explanation. Why should God have thus conditioned man as an elaborate copy of the ape, when we know from the rest of creation how endless are His resources in the invention of types?

—GEORGE J. ROMANES, 1882*

Evolution wouldn't be such a controversial subject if it didn't touch on our perceptions of ourselves. In the Western tradition, humans are set apart from the natural world. The gap in mental and emotional

---

* George J. Romanes, a defender of Darwin, was secretary of the Linnaean Society of London, the major scientific organization of the day.

powers between humans and animals is thought to be a profound, unbridgeable difference in kind. According to this anthropocentric, even egocentric tradition, the earth and its inhabitants were created to serve us. We are the special object of God's creative beneficence, so much so that he will even bend the natural world to our desires, and alter natural laws in response to our special pleading. Nothing could be more antithetical to such a world view than a science that tells us the earth is not the center of the universe; that life came and went for billions of years before man appeared on the scene; that living things and the human species itself originated by natural, impersonal causes rather than the direct intervention of a Creator; that we are as much a part of nature as each of the millions of other species with which we share a common bond of inheritance.

Long before Darwin and Wallace, biologists were well aware of the similarities between humans and apes. Thus Linnaeus remarked in a letter in 1747, "I demand of you, and of the whole world, that you show me a generic character, by which to distinguish between Man and Ape. I myself most assuredly know of none. I wish somebody would indicate one to me. But if I had called man an ape, or vice versa, I should have fallen under the ban of all the ecclesiastics. It may be that as a naturalist I ought to have done so."[1] In the early nineteenth century, European biologists came to believe that the non-Caucasian races were unchanging links in the great chain of being, between apes and the "advanced" races of Europeans. The scientists of that day were as thoroughly imbued with racism as the rest of the society in which they moved. They assumed that the cultural differences between "savages" and the exalted race to which they belonged were hereditarily fixed.

Nevertheless, by the beginning of the nineteenth century, the notion of cultural change had started to take hold. Rousseau, Lord Monboddo, and Cuvier all speculated that humans had progressed (not evolved) from an initial state of savagery to higher and higher levels of civilization. The idea that the physical, biological characteristics of the human race might have changed did not come until later; but the way for biological evolution was paved by the concept of cultural change. If human culture could progress, why not the human race itself? It was quite clear to Darwin that the publication of such a hypothesis was not to be taken lightly. The protest that greeted *The Origin of Species* had been so predictable that Darwin carefully avoided any discussion of human evolution in that book. He limited his comment on the topic to one pregnant sentence, "Light will be thrown on the origin of man and his history," at the end of the *Origin*. Twelve years passed before

Darwin directly addressed the subject, in *The Descent of Man, and Selection in Relation to Sex*. By this time, the first furor over evolution had passed, and evolution was widely enough accepted among scientists that Darwin could hope for some acceptance of his views on humans. These were, chiefly, that humans and apes had developed from a common ancestor; that the human intellect and emotions are greatly magnified compared to those of other animals, but do not differ in kind; that like our physical characteristics, they evolved by natural selection; and that many of the differences among races, as well as many of the physical features peculiar to humankind, were the result of a special form of natural selection that he called "sexual selection." Sexual selection, to which he devoted more than two thirds of the book, results in the evolution of features that do not promote survival, but merely give their possessors an advantage in reproduction, by enabling them to acquire mates.*

Since Darwin's day, we have acquired a fossil record of human evolution that grows richer each year. We have also amassed evidence of our affinity with apes of which Darwin could not have conceived, and have learned so much about the behavior of apes that it has become almost impossible to define any aspect of human behavior as being truly unique.

Most taxonomists place humans in one family, the Hominidae, and the chimpanzee, gorilla, and orangutan in another family, the Pongidae. Some feel, however, that there is no reason to classify humans as a separate family. Each of the apes has its own specialized features, and surely all have evolved since the hominid and ape lineages separated millions of years ago. Hence we do not expect our common ancestor to be exactly halfway between humans and chimpanzees, any more than the common ancestor of horses and rhinoceroses was half horse and half rhinoceros. Despite the peculiarities of each species, however, the anatomical similarities between humans and apes are striking. Apes and humans can be matched bone for bone and muscle for muscle. Most of these elements differ in size and shape, as they do among the various species of apes, but they are all there. As Darwin pointed out, humans share with apes vestigial features that are clearly homologous (derived from a common ancestor) with those of other mammals, such as now useless muscles that once moved the ears and tail, and the vertebrae of the tail itself. We have muscles that erect the hairs of our body when

---

*See pages 121–22 for a more detailed discussion of sexual selection.

we are cold or afraid, even though this no longer does us any good. (In longer-haired mammals, erection of the hair increases insulation, and, as in cats, makes the animal look larger when it faces an enemy.)

Many of the modifications of human bones and muscles are adaptations to our bipedal (two-footed), erect posture, such as the shape of the feet, pelvis, and vertebrae, and the position of the juncture between skull and vertebral column.[2] Very pronounced differences also exist in the head. Our skull is greatly enlarged relative to the size of our body, because it houses a much larger brain (average 1,400 cubic centimeters, or cm³) than that of the apes (400 cm³ in chimpanzees, 500 cm³ in gorillas). Our cranium is more globular, and our forehead rises directly above the front of our face, which is greatly foreshortened compared to that of apes, with their protruding snouts. Our cheek teeth are set in a parabolic arch rather than parallel; the incisors are more shovel-shaped, the canines much smaller, and the cheek teeth smaller and more flattened.

Our greatest distinction is obviously in behavior rather than anatomy; but for all our glorious abilities, anthropologists have had a hard time defining just what is uniquely human in our behavior. At one time we were supposed to be the only tool-using species; but many primates, and even other animals such as sea otters and finches, use tools. Then we were supposed to be the only species that *makes* tools; but Jane Goodall discovered that chimpanzees fashion sticks to probe into termite nests, and that this behavior is passed on from generation to generation by learning. In fact, quite a few primates have "cultural traditions." Troops of Japanese monkeys, for example, have developed a tradition of tossing sand into water to separate out seeds that they eat.

One of our distinctive features is supposed to be our self-image: we are conscious of ourselves as individuals. It is hard to know if other species have such consciousness, but chimpanzees primp in front of mirrors to adjust their appearance, suggesting that they are conscious of what they "should" look like.[3] The big issue, though, is whether or not the human capacity for language is unique. It was always believed that this set us off from all other species. In recent years, however, even this was brought into question, because several psychologists started to teach chimpanzees to use sign language. Although such experiments are still highly controversial, a number of psychologists have come to believe that chimpanzees can use signs as true language: they seem to generalize, using an arbitrary sign for both a specific object and the larger class of objects to which it belongs; they combine signs in new combinations with a rudimentary syntax, and on occasion they even seem to invent new symbols. Whether or not apes can "talk," it is clear that the genetic capacity for learning and rudimentary culture is present in apes and even other animals, though to a lesser extent by far than in humans. Much of psychology uses animals as experimental subjects for studying learning, perception, and the functioning of the brain, with the aim of understanding the mechanisms of human behavior; if such mechanisms were not fundamentally the same, this entire approach would be barren.

Some of the anatomical peculiarities of humans aren't quite as peculiar if viewed in an embryological perspective. Many authors have pointed out that humans might be "neotenic"—that is, that they retain juvenile features into adulthood.[4] In many ways, we look like an enlarged version of a human—or ape—fetus. Our foreshortened face, nonopposable big toe, globular skull, and relatively large brain are just a few of the features you would expect if a primate fetus got bigger but didn't "grow up." Certainly not all our characteristics can be explained this way, but it is tempting to suppose that many of the anatomical differences between apes and humans could be due to only a few genes that control the relative growth rates of different parts of the body. Perhaps the genetic differences between apes and humans aren't as great as their adult anatomy suggests at first.

This possibility is reinforced by recent data from molecular biology, which provide the most extraordinary confirmation of our relationship to apes. We must step for a moment into molecular genetics to appreciate these data.

A gene is a length of DNA, made up of a specific sequence of nucleotide bases, which come in four kinds: adenine (A), thymine (T), cytosine (C), and guanine (G). Many genes contain coded instructions for making proteins, which are linear chains containing up to twenty kinds of amino acids. Each triplet of nucleotide bases in the DNA corresponds to one kind of amino acid: thus the DNA sequence AAA codes for the amino acid phenylalanine, and CTC codes for glutamic acid. A change in a nucleotide, which then is propagated into future generations when the DNA replicates itself, is a mutation. If, for example, the triplet CTC mutates into CGC, there will be a hereditary change in the protein molecule that the gene codes for: alanine will replace glutamic acid at a specific place in the protein chain. An average protein contains about 500 amino acids, corresponding to 1,500 DNA nucleotides, so many different mutations of the same protein are possible.

The number of amino acid differences between two species in one

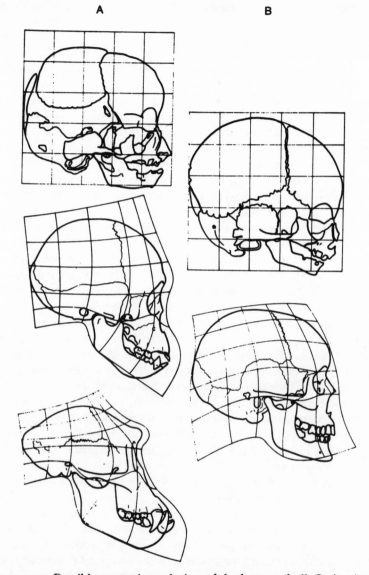

FIGURE 19. Possible neotenic evolution of the human skull. Series A, at left, shows the growth of the skull of a chimpanzee; series B, at right, that of a human. The intersections of the grid mark the corresponding points on the skull as development proceeds. The grid becomes less distorted in human than in chimpanzee development, showing that skull shape changes less from the fetal form in humans than in chimpanzees. (From D. Starck and B. Kummer, *Anthropologischer Anzeiger* 25:204–15 [1962].)

of their proteins is considered a good measure of how genetically different the species are. In general, the degree of genetic difference measured
in this way corresponds very well to the length of time since the species
are believed to have diverged from their common ancestors. Various
mammals, for example, are more similar to each other than they are
to reptiles, and more similar to reptiles than they are to fishes. Over
very long periods of evolutionary time, the average rate of divergence in DNA appears to be roughly constant. Many biologists therefore believe that differences in protein or DNA can be used as a kind
of "clock" that will tell when species separated from their common
ancestors.

The similarities of ape and human proteins are extraordinary.[5] For
example, hemoglobin, the oxygen-carrying protein in the blood, contains 287 amino acids in an identical sequence in chimpanzees and humans. Compare this to the difference between two closely related
species of frogs, whose hemoglobin differs in 29 amino acids. Out of
the 153 amino acids in myoglobin, a muscle protein, humans and chimpanzees differ in only one. Mary-Claire King and Allan Wilson, biochemists at the University of California at Berkeley, estimate from an
analysis of 12 kinds of proteins that chimps and humans differ on average at only 7 out of every 1,000 amino acids. The evidence from proteins and from direct analysis of DNA indicates that humans and apes
are more genetically similar to each other than certain species of fruit
flies or rodents that are *identical* in external appearance. Apes and
humans are so similar that Allan Wilson and his colleague Vincent
Sarich have suggested[6] that the human lineage branched off from the
gorilla-chimpanzee line only about 4 to 5 million years ago, and not
the fifteen or so million years previously estimated on the basis of the
very inadequate fossil record.

The same conclusion, of genetic near-identity, comes out of a recent detailed analysis of chromosomes, the structures that carry the
genes.[7] The chimpanzee has 24 kinds of chromosomes, humans have
23. (Each chromosome type comes in pairs, so the total number is 48
for chimpanzees and 46 for humans.) The chromosomes are marked
by multitudes of narrow dark bands that can be matched perfectly
against each other in the two species. The only major differences are
in the arrangement of certain segments of some of the chromosomes,
which in the chimpanzee are inverted 180 degrees from their arrangement in humans, and in the fusion of two of the chimpanzee chromosomes into one in the human set. Such differences are minor
compared to chromosome differences that are often found among

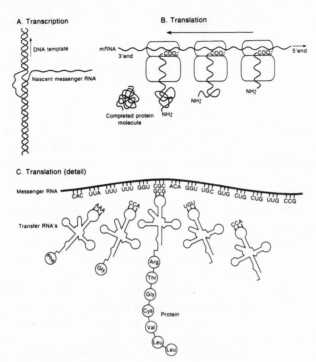

FIGURE 20. The steps leading to protein synthesis. In *A* (Transcription), the two spiral helices of a DNA molecule separate. One half of this molecule consists of a series of nucleotides, adenine (A), thymine (T), cytosine (C), and guanine (G). Corresponding to these, a series of nucleotides are strung together to make a messenger RNA (ribonucleic acid) molecule, which has uracil (U) wherever the DNA has adenine, adenine (A) where DNA has thymine (T), G where DNA has C, and C where DNA has G. Protein molecules are built according to the RNA code at assembly sites called ribosomes, in the process of "translation" *(B)*. In this process *(C)*, specific transfer RNA molecules carry amino acids (e.g., phenylalanine, Phe, and glycine, Gly) to sites on the messenger RNA determined by the correspondence between a triplet of nucleotides (e.g., CGC) on the messenger RNA and a triplet (e.g., GCG) on the transfer RNA. The amino acid carried by the transfer RNA molecule is added to the growing protein chain. Thus a DNA sequence TGT-GCG-ACC-AAA is transcribed into the messenger RNA sequence ACA-CGC-UGG-UUU, which in turn is translated into the amino acid sequence threonine (Thr)–Arginine (Arg)–Glycine (Gly)–Phenylalanine (Phe). (From *Evolution* by Theodosius Dobzhansky, Francisco J. Ayala, G. Ledyard Stebbins, and James W. Valentine. W. H. Freeman and Company. Copyright © 1977.)

related species of plants or rodents, and probably do not have much of a genetic effect.

Thus even without considering the fossil record, the more closely we look, the more evidence we find of our genetic relationship to the apes. In the last few decades, though, our understanding of human evolution has been greatly enhanced by a series of spectacular paleonto- logical discoveries.

Before tracing the fossil record in detail, we should consider what we might expect to find if human evolution were like that of other species. First, at any given time there will be variation within each population, and average differences among populations, just as there are today. Second, different characteristics will evolve at different rates, and may evolve within one population of the species, but not others, in response to local environmental conditions. Therefore it is possible for some populations to retain "primitive" characteristics after others have evolved to a new form. It is even possible for evolution to become "reversed," since it is not an inexorable march to some predestined goal. Third, evolution need not proceed at a constant, steady rate; it may be faster at some times than at others. Finally, an ancestral species is likely to give rise to several descendant species, of which some may become extinct while others will survive and change. Thus not every hominid fossil is necessarily on the direct line leading to modern humans. Never- theless, it should be possible, with a complete enough record, to trace transitions back through various intermediate stages to more apelike ancestors.

The literature on human paleontology can be very difficult to read. Until recently, most paleontologists concerned with human evolution knew little about the genetic theory of evolution. Partly for this reason, and partly because of their desire for fame and reputation, anthropolo- gists have had a history of giving each new fossil a separate name, as if it were an unprecedented discovery of such importance that it merited advertising. Most of these names ("*Paranthropus,*" "*Megalanthropus,*" "*Zinjanthropus,*" and so on) are no longer used. The generic names now used for hominid fossils are simply *Australopithecus* and *Homo,* and even the distinction between them is arbitrary. Moreover, even though differ- ent kinds of *Homo* are still given different names (*Homo habilis, Homo erectus, Homo sapiens*), these aren't distinct either. One "species" blends into another as you go from older to younger fossils.

The oldest relevant fossils are apelike forms called dryopithecines that include the genus *Ramapithecus,* which comes from Pliocene deposits in India that are 10 to 14 million years old.[8] The anthropologists David Pilbeam and Elwyn Simons, who have studied these fossils carefully, once argued that *Ramapithecus* may be on the human line. If so, humans branched off from the line leading to modern apes more than 10 million years ago. However, the fossils are only fragments of skulls and jaws, and they do not have any uniquely, indisputably hominid characteristics. It may turn out that *Ramapithecus* was a common ancestor of both the Hominidae and the Pongidae. If so, then the split between the pongid and hominid lines could have occurred after *Ramapithecus,* as the biochemical similarity between apes and humans seems to suggest.

The first really useful fossils come from the late Pliocene, 3.7 to 3.5 million years ago, when Africa had extensive grasslands. Recently, Donald Johanson, Richard Leakey, and other paleontologists have found abundant Pliocene fossils in Ethiopia and Tanzania of ground-dwelling primates that are almost certainly on the hominid line leading to modern humans. These include the spectacular skeleton that Johanson and the anthropologist Theodore White have dubbed "Lucy" and given the scientific name *Australopithecus afarensis.*[9] This species was clearly apelike in many respects. It had relatively long arms and short legs, curved finger bones, and a parallel dental arch. But the pelvis and leg bones show it to be almost fully human in one crucial respect: it was bipedal. "Lucy" stood fully erect, at about four feet. Her skull was probably small, but the fossils are not complete enough to indicate the precise size of the brain.

The first australopithecines were found in South Africa by Raymond Dart in the 1920s. Unlike the Tanzanian fossils, they unfortunately cannot be dated by radiometric methods, but other geological evidence indicates that they are at most 2 to 3 million years old. They include two forms that almost certainly were separate species adapted for different ways of life. *Australopithecus robustus* was a stocky creature, with a 530 cm^3 brain and very massive jaws and teeth that seem clearly adapted for a diet of seeds and other plant materials. Remains of *robustus* persist in the Olduvai Gorge of Tanzania until about 1.5 million years ago. The Tanzanian fossils of *robustus* were described by the late Louis Leakey (Richard's father), who had an extraordinary knack for finding human fossils. Leakey named them *Zinjanthropus boisei,* but it is clear that they are merely a more robust version of the South African *robustus.* The most recent fossils of *robustus* are about 1 million years old, after which the species seems to have become extinct.

Coexisting with *Australopithecus robustus* in South Africa there was a more slender, probably omnivorous species that Dart named *Australopithecus africanus*. It was quite similar in many ways to *afarensis*, the name given to "Lucy," and many anthropologists feel that they do not deserve separate names. *Africanus*, like *afarensis*, was fully bipedal, but it also had some other more human characteristics such as a short canine tooth and a parabolic dental arch. It is found in association with "pebble tools" that it apparently made by knocking chips off larger stones. *Australopithecus africanus* had a brain size of about 440 cm³; thus it combined hominid features with an ape-sized brain. As in evolution generally, different human characteristics evolved at different rates. Our erect posture preceded the enlargement of our brain.

Among the many fossils that Louis Leakey and his wife Mary discovered in the Olduvai Gorge in Tanzania were several that he named *Homo habilis*. These can be dated through the early Pleistocene from 2 to 1.6 million years ago. In fact, the differences between *Homo habilis* and *Australopithecus africanus* are so slight that many anthropologists feel the distinction between the two is totally artificial. The chief reason for dignifying *habilis* with the name *Homo* is that it had a larger brain: about 600 cm³ rather than 440 as in *africanus*. *Habilis* is associated with an extensive "industry" of pebble tools, which become increasingly sophisticated in the upper Olduvai beds. Several characteristics of their teeth also change gradually toward a more modern, human state.

The next stage in human evolution is represented by some of the first human fossils that were ever found. The "Java man," discovered by Dubois in the 1890s, was named *Pithecanthropus erectus*, the "erect ape-man," and was essentially a modern human in every aspect except its brain size, which ranged from 750 to 900 cm³. Subsequently, many specimens of Java man were found in China, but unfortunately they were lost during World War II. All that remains are plaster casts. The same species, now known as *Homo erectus*, is known from Africa, however, where its remains range over a geological span from 1.6 to about 1.3 million years ago. *Erectus* is essentially the same as *habilis*, except for its larger brain, which in the African fossils ranges from 850 to 1,000 cm³, and is larger in more recent than in older specimens. It also is associated with more sophisticated tools than *habilis*, including stone hand axes.

Homo erectus was essentially modern in size and posture, and almost fully modern in dental characteristics. The face still projected in a somewhat apelike form, but was flatter than in *habilis*, and the cranium, although larger, still had a low, sloping forehead. The Chinese skulls, however, which are estimated to be 800,000 to 500,000 years old, appar-

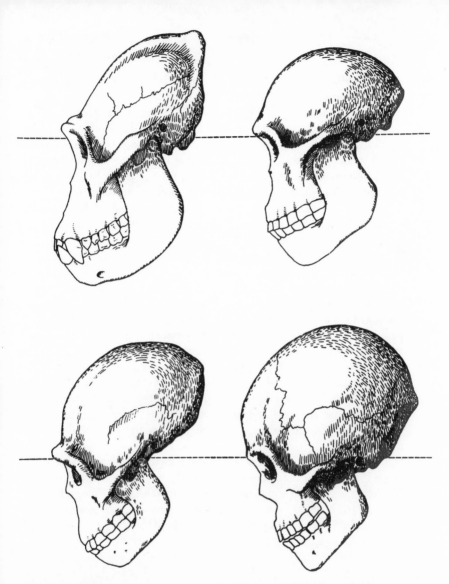

FIGURE 21. The skull of a modern ape (upper left); *Australopithecus* (upper right); *Homo erectus* (lower left); and a modern human (lower right). The horizontal line separates the face and jaw, which are progressively reduced in the hominids, from the brain case, which is progressively enlarged. The inclusion of the ape skull is *not* meant to indicate that *Australopithecus* emerged from a species that looked like this. (From "Tools and Human Evolution" by Sherwood L. Washburn. Copyright © [1960] by Scientific American, Inc. All rights reserved.)

ently had a more modern shape, and ranged in brain size up to 1,300 cm³. There is also evidence that *erectus* used fire.

Thus by about 1.5 million years ago, *Homo* had expanded out of Africa throughout the tropical and subtropical Old World, and had a fairly advanced culture. By the late Pleistocene, 200,000 to 100,000 years ago, humans of almost fully modern form had expanded into Europe. Their brain size was on average 1,200 cm³, as large as that of many living humans, although not quite up to the modern average (1,400 cm³). Whether the fossils from this period should be considered *Homo erectus* or *Homo sapiens* is entirely arbitrary, especially because from 200,000 years onwards there was a rapid change to the fully modern condition: the skull became rounder, the face, teeth, and brows were reduced. By the end of the Pleistocene, 75,000 to 45,000 years ago, the brain had reached its modern level. In the Near East, people had 1,400 cm³ brains; in Western Europe they had even larger brains. These western populations, known as Neanderthals, were made up of massive, heavy-browed people with brains averaging 1,600 cm³—rather larger than the modern average. They had a sophisticated industry of stone tools, and doubtless were thoroughly intelligent people. Between 30,000 and 50,000 years ago, however, the Neanderthals disappeared utterly from the fossil record. Why this should be so is a matter of speculation. It has been suggested that they suffered in warfare with smaller-brained people moving in from the east, or lost their identity by interbreeding with them.

With the introduction of agriculture about 12,000 years ago in the Near East, we reach the scene that ends this strange eventful history. The stage was set for a new drama: modern civilization. From this time on there has been, as far as we know, no appreciable physical or mental evolution of the human species. A new form of "evolution," nongenetic cultural change, now shapes our ends.

It is only in the last few decades that most of this fossil evidence has come to light. Now that paleontologists know where to look, it seems likely that the gaps in the record will be steadily filled in. The hominid fossil record is far from complete. We don't by any means have a fossil for every thousand years of history, but what we do have shows a remarkably steady evolutionary movement.[10] Our bipedal posture, which defines the australopithecines as hominids, evolved first, and was improved with the passage of time. Dental characteristics, too, show steady movement toward modern form. We don't know if brain size evolved continuously, but it looks as if it progressed steadily, *on average*. When average brain size is plotted against time, it shows a remarkably straight line. It went from 440 cm³ 2.8 million years ago *(africanus)* to 600 cm³ at 2 million years *(habilis)* to about 930 cm³ at 1.1 million years *(erectus)* to 1,400 cm³ and 1,600 cm³ (Neanderthals) by the time we reach *sapiens,* half a million years ago.

Of course, there is no reason to think that all this evolution happened at an utterly steady pace; indeed, it did not. Brain size increased on average at a constant rate, but around this average there were undoubtedly fluctuations, short-term accelerations and decelerations of evolution. Adherents of the hypothesis of punctuated equilibrium, in fact, have emphasized these fluctuations in rate, and have suggested that the accelerations happened when local populations became new species, evolved larger brains, and replaced the more slowly evolving species from which they had diverged. The fossil record is still not complete enough to describe the process in great enough detail to evaluate this hypothesis, but it does give as nice a series of ancestors and descendants as the horses do. Indeed, the hominid fossil record is rapidly becoming one of the best examples of microevolution.

Just what the forces of natural selection were that directed this evolution, and what implications our primate ancestry has for modern human behavior, are topics that have provoked an enormous amount of often quite irresponsible speculation. It is certainly possible to imagine reasons for our peculiar evolution, but probably impossible to subject any of these ideas to serious scientific testing. For example, we may suppose that as savannas replaced forests during the Pliocene, the ability to run was more advantageous than the ability to climb trees, and that this would favor the bipedal habit. If troops of australopithecines developed the habit of hunting for animals, there would be an advantage to cooperative behavior, which would favor the development of a complex mind that could promote social alliances, lay ambushes for prey, and lead migrations to favorable sites of food and water. The more complex the social interactions, the more beneficial still greater intelligence would be. At some point, the ability to identify other individuals and plan social activities might have required so complex a brain that it became capable of self-awareness and language, and with them, the capacity for fantasy, artistry, and rational thought.

But this is only one of many plausible stories. We could as well adopt Robert Ardrey's view in *The Territorial Imperative,* [11] and suppose that aggression and warfare among tribes of australopithecines was the prime mover behind the evolution of consciousness and intelligence.

But all such stories, I feel, remain speculations, for the fossil record offers no direct evidence on what factors caused human evolution to occur. It only documents that as we go back in time, the physical characteristics of hominids become more apelike, and the cultural artifacts such as stone tools become simpler in form, just as we would expect if humans are the product of a gradual evolution of physical and mental traits.

The paramountly important feature of human evolution is the development of consciousness and its various manifestations: language and culture. These abilities have radically altered the course of human evolution and have brought many (but not all) aspects of human evolution almost to a halt. We may still be genetically evolving adaptations—for instance, to diseases and air pollution—but culture enables us to solve many adaptive problems without genetic change: environmental factors such as cold weather and predators cease to select some genotypes* over others if everyone is protected by housing, weapons, and fire. As far as we can tell, the human capacity for culture arrived some tens of thousands of years ago at its modern level, and there has been little if any genetic change in our mental abilities since then. Human populations differ genetically in trivial ways such as skin color and blood types, but there is no evidence whatever that they differ genetically in mental abilities.[12] Nor is there any reason to think that changes in culture in the last 50,000 years or so have been caused by any changes in the genes. In humans, the rudimentary primate capacity for learning cultural traditions has become greatly exaggerated, so that our behavior is now influenced primarily by a new form of inheritance: the passage of knowledge and cultural traditions from generation to generation. Our cultural inheritance can undergo immense changes within a single generation, so that social "evolution" is now inconceivably more rapid than

---

* The phenotype is the observable condition of some characteristic of an individual organism: whether a person has hemophilia or not, brown eyes or blue, good musical ability or poor. The genotype is the constitution of genes the individual has that affect the characteristic. The relation between genotype and phenotype is sometimes a simple one-to-one correspondence, as in the case of human M-N blood types. Very often, however, it is much more complex. A single genotype can produce many different phenotypes, depending on environmental conditions. Thus aphids with the same genotype may either develop wings or not, depending on how crowded they are. Or a particular phenotype may be produced by many different genotypes. The simplest example (by far) is provided by dominant and recessive genes. Thus a person with blood type A may have either the genotype AA or the genotype AO, whereas two O alleles (genotype OO) produce the blood type O.

genetic evolution. Thus the historian, the archaeologist, and the linguist trace changes in social structures, tools, agricultural methods, religious beliefs, and languages, none of which have changed genetically.

However, the process of cultural evolution is almost as much anathema to the creationists as the process of biological evolution. The fossil evidence of our physical evolution is, of course, not acceptable to them. As far as they are concerned, *Australopithecus* and *Homo erectus* were apes, not human ancestors—despite the fact that if *Australopithecus africanus* wasn't our ancestor, it is remarkably similar to what we would expect our ancestor to look like. But the creationists feel that "it is in the realm of the social sciences that the difference between evolutionist and creationist philosophy is most important."[13] Thus different languages couldn't have evolved from an original language, they say, because "primitive" tribes have complex languages. Instead, they suppose that, after the worldwide flood, the Creator directly restructured a primal language into the diverse languages that exist today. All the thousands of genetically and culturally diverse peoples of the world, they claim, originated from the survivors of that flood in the vicinity of Mount Ararat in the Near East, which thus becomes the center from which all civilization spread.[14] The differences between races and tribes arose when the Creator imposed different languages on the survivors of the flood, which impelled them to separate into different groups. But, they say, this idea is not susceptible to scientific investigation.[15]

All of anthropology militates against such a view. The complex subject of cultural anthropology, however, is not one in which I am qualified. It is important to recognize, though, that the creationist attack is not limited to biology: the social sciences are next in line for Biblical reinterpretation.

SIX

NATURAL
SELECTION
AND ADAPTATION

N othing in biology makes sense except in the
light of evolution.
—THEODOSIUS DOBZHANSKY, 1973

On the cover of a recent issue of *Awake!*, a publication of the
Watchtower Bible and Tract Society, a child looks a turtle in the eye.
"Accidents of Evolution? or Acts of Creation?" Inside, an article enti-
tled "Design Requires a Designer" asks, How is it that "evolutionists
can blithely assign to chance the power to design all complex living
creatures?"

Strong is the power of the well-chosen word. "Accidents" and
"chance" certainly could not suffice to produce the order and complexity of living things, but no biologist claims that they can. "Design" does
indeed imply a "designer"; for a design is a plan or scheme. However,
where a creationist sees a design or plan, a scientist sees merely order,
or regular arrangement, and order does not require a designer. In fact,
one of the chief tasks of science is to determine how order can be
produced by inanimate natural forces.

The fact is, order in nature is no evidence of design. Visit a beach
and look at the arrangement of pebbles and sand grains. They are not

randomly dispersed: the larger pebbles are farthest from the waves, sifted into a regular arrangement by the force of the water. Open a chemistry book and find that a crystal of salt has a perfect geometrical order because one electron on each sodium atom fits nicely into a corresponding space in the orbital of each atom of chlorine. Consider a meteorite: it no longer sweeps around the sun in a regular orbit because it came too close to the earth. Can anyone doubt that the asteroids and planets continue on their paths merely because they are too far away to attract each other into colliding? They form a stable, ordered system because they are the only bodies in the solar system that are left, after gravitational forces sorted out the unstable orbits from the stable ones.

The great triumph of Newtonian physics was that it was able to work from a few basic principles that could be observed in everyday life to an explanation of the visible universe. The same natural laws that explained the trajectory of a cannon ball could be extrapolated to the motion of the heavenly bodies. An astronomer cannot experiment with the planets. He or she can only observe whether the orderly pattern of the planets is consistent with principles that can be tested by experiments here on earth—whether dropping objects from the Leaning Tower of Pisa or splitting atomic particles in a nuclear accelerator. Thus much of science progresses by explaining large phenomena in terms of particular mechanisms that can be observed here and now.

Biological systems, like nonliving systems, are ordered—and like nonliving systems, they can be explained in terms of detailed physical mechanisms. Physiologists may be daunted at times in attempting to understand how a cell carries out the intricacies of metabolism; but as physiology has grown, it has shown these intricacies to be the consequence of fairly simple laws of chemistry, not of a mysterious "life force." Ecologists wonder why different kinds of soil should consistently support different species of trees, instead of a jumble of all possible species; but they then find that competition between plants relegates each species to the kind of soil where it has an advantage over its competitors. The key to understanding in each case is not to throw up one's hands in despair when faced with complexity, but to experiment with isolated parts of the complex system and see if the insights gained from the particular will serve to explain the whole.

If, then, there is some natural mechanism that can produce the wonderful order that every living thing embodies, we should be able to observe this mechanism acting, and do experiments that will show how it acts. This mechanism is the antithesis of chance. It is a process that forms order out of disorder in the living world, just as the laws of physics create order among the pebbles on a beach.

Consider a simple experiment that has been done many times. Put two strains of bacteria in a flask with a continuously stirred nutritious broth and replace some of the broth every day with new broth. Both kinds of bacteria multiply, and both are removed at a constant rate when the broth is replaced. If for some reason one strain of bacteria can feed faster and thus divide more rapidly, it will make up a greater and greater percentage of the total bacterial population, and ultimately will reach 100 percent. There is really nothing more to natural selection than this: if, in a particular environment, one kind of organism reproduces faster or dies out more slowly than another, it will tend to replace the slower-growing form. In order for this to happen, there must be some initial variation. Perhaps the faster-dividing strain of bacteria has an enzyme that metabolizes sugar more rapidly. The variation, moreover, must be hereditary. The capacity for more rapid sugar metabolism must be passed on from parent to offspring. Natural selection, therefore, is merely a name for any consistent difference in survival or reproduction between genetically different members of a species.

An essential point in this definition is that the difference be *consistent*. Whenever these two strains of bacteria are in an environment where sugar is in short supply, the one will consistently take over. The outcome of competition between the bacteria isn't a matter of chance; it is a predictable consequence of the difference in their biochemical capacities. Chance may well dictate whether these bacteria find themselves in an environment that is low in sugar; chance may well determine whether a population of bacteria contains a genetic mutant that can metabolize sugar more rapidly; but if what a mathematician would call the initial conditions exist—if there is little sugar, and an efficiently metabolizing genotype of bacteria is present—then the more efficient genotype will predictably replace the other.

A common consequence of natural selection is adaptation, a more effective means by which the organism utilizes its environment. The bacteria in the experiment may be said to become better adapted to a low-sugar environment by replacing less efficient genotypes with a more efficient one. We will not expect adaptation to be perfect, of course. The genotype that takes over isn't necessarily the most efficient possible one, which may not have arisen in the bacterial population; it is just the most efficient of those that were present. Nevertheless, the outcome of this process will be a closer match between the bacterial enzymes and the environment in which they function, a match that will look as though it had been planned or designed.

Charles Lyell, the geologist who was one of Darwin's closest friends, likened natural selection to the Hindu deity who has three faces: Shiva the destroyer, Vishnu the preserver, and Brahma the creator. As Shiva, natural selection relentlessly destroys the unfit as they appear: innumerable harmful mutations arise constantly in every population and are eliminated when they fail to survive or reproduce. As Vishnu, natural selection tends to hold a species to the status quo: if a certain body size is advantageous, both smaller and larger individuals will be less successful, and so genes that deviate in either direction will be eliminated. As Brahma, natural selection may favor new characteristics—better sugar-metabolizing ability or larger size—and act as a creative force, shifting the species toward a new optimum state.

When Darwin wrote *The Origin of Species*, he could offer no good cases of natural selection, because no one had looked for them. He drew instead an analogy with the artificial selection that animal and plant breeders use to improve domesticated varieties of animals and plants. By breeding only from the wooliest sheep, the most fertile chickens, and so on, breeders have been spectacularly successful in altering almost every imaginable characteristic of our domesticated animals and plants to the point where most of them differ from their wild ancestors far more than related species differ from each other.

Since Darwin's time, geneticists have carried out many experiments in the laboratory, and have shown that almost any species can be rapidly altered well beyond its original range of variation. A common experiment, for example, is to select fruit flies for greater numbers of the bristles that are situated on the side of the body, merely because these are easy to count. In a recent issue of *Genetical Research*, B. H. Yoo[1] reports that a population of flies increased in bristle number continually for about ninety generations before the experiment was arbitrarily ended. Initially, most of the flies had eight to eleven bristles. By the end of the experiment, the average fly had thirty-six bristles.

Experimenters have also selected species for truly novel, adaptively important characteristics. A British microbiologist, Patricia Clark,[2] has selected strains of *Pseudomonas* bacteria that can grow on the chemical phenylacetamide, which that bacterium normally cannot metabolize. This capacity of bacteria for biochemical evolution plays an important role in industrial chemistry, in which new strains of bacteria are selected to carry out the synthesis and degradation of many organic molecules. Evolution in the laboratory has also been observed many times in insects that are exposed to insecticides. Expose a population of flies to DDT, malathion, or almost any other poison, and within a few generations the population has evolved to be resistant.[3]

But does natural selection occur in nature, without the guiding influence of a geneticist or plant breeder? Indeed it does. It is going on all around us, and even within us. One of the most serious drawbacks of using antibiotics too freely is that bacteria evolve resistance to them within our own bodies. The more antibiotics we use, the more powerfully we are favoring mutant forms that resist it. This is one of the major reasons why medical researchers are continually searching for new antibiotics: they are in a race to stay ahead of bacterial evolution.

Chemical companies, too, are racing to keep ahead of insects' abilities to evolve resistance to insecticides. In less than thirty years, more than two hundred species of insects have evolved resistance to the DDT

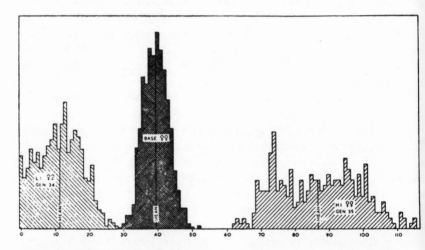

FIGURE 22. The results of thirty-five generations of selection in the laboratory for changes in the number of bristles in fruit flies. The original (base) population had an average of 39 bristles, and ranged from about 29 to 52. Two subpopulations derived from these flies were selected for either higher or lower bristle number. The "high" subpopulation (right) had an average of 87, and the "low" subpopulation (left) an average of 11 bristles, after thirty-five generations. In both subpopulations the range of variation was shifted completely out of the bounds observed in the original base population. (From G. Clayton and A. Robertson, *J. Genet.* 55:154 [1957].)

used to protect crop plants and control disease-bearing mosquitoes. Another way to combat insect pests is to breed insect-resistant plants, but the insects can and do evolve to counteract this tactic, too. For example, strains of wheat resistant to a major pest, the Hessian fly, were developed. However, new strains of Hessian fly have more than kept pace, forcing the development of still more resistant types of wheat.[4] One of the major tasks of the International Rice Research Institute in the Philippines is to breed new, resistant strains of rice faster than the brown plant-hopper can evolve to destroy the major food crop of eastern Asia.

Just as Darwin had suspected, even a relatively slight difference can sometimes have an important impact on survival, and thus result in evolution. On February 1, 1898, for example, a severe storm battered eastern Massachusetts and left hundreds of dead and dying birds in its wake. Someone brought 136 exhausted sparrows to Hermon Bumpus, a professor at Brown University.[5] About half the birds promptly died, but the others regained their health. Bumpus killed them, however, so that he could measure their skeletons. He found that among the male sparrows, the larger birds had survived better than the average or small ones. What is most remarkable was that even a small percentage difference in size was enough to influence survival. The average length of the breastbone of the surviving males was 21.76 millimeters, compared to 21.46 in the non-survivors—yet this was enough to affect their survival.

A rather similar observation, but with more details, was reported in the October 1981 issue of *Science* by Peter Boag and Peter Grant.[6] They have been studying the Darwin's finches of the Galápagos Islands for about ten years, and found that a drastic change in the environment has altered the genetic composition of one of the species of finches quite appreciably. During 1977 the islands suffered a severe drought, so that there was a pronounced drop in the production of the small seeds on which the finches feed. The finches were forced to feed on larger seeds, which they usually ignore. After one generation, there had been so much mortality of the smaller birds, which could not feed efficiently on large seeds, that the average size of the birds, and especially the size of their beaks, went up appreciably. Very possibly the birds will evolve back to their previous state if the environment goes back to normal, but we can see in this example what would happen if the birds were forced to live in a consistently dry environment: they would evolve a permanent adaptation to whatever kinds of seeds are consistently available. This is natural selection in action, and it is not a matter of chance.

One well-known example of natural selection is sickle-cell hemoglobin in human populations in Africa. This form of hemoglobin, called hemoglobin-S, differs from "normal" hemoglobin (hemoglobin-A) at just one position in the protein molecule, where the amino acid valine is substituted for glutamic acid. This alteration of the molecule reduces the ability of the red blood cells to carry oxygen. As for almost any gene, a person may be homozygous, meaning that she or he has inherited the same form of the gene—the same allele—from both parents; or she (he) may be heterozygous, having inherited a different allele from each parent. Thus each person has two representatives of a gene, and the representatives may be the same or different alleles. Hemoglobin-S and hemoglobin-A are produced by different alleles of the gene that codes for one part of the hemoglobin molecule.

People who are homozygous for the sickle-cell allele have only hemoglobin-S. They are severely anemic and usually die before they can grow up to reproduce. People who inherit one sickle-cell gene and one "normal" gene are heterozygous, and although they are slightly anemic they usually survive. Individuals who have only hemoglobin-A (are homozygous for the A gene) are not anemic, but in malaria-infested regions of Africa they suffer malaria more than heterozygous people do, and often die from it. Therefore there is a balance: both the A and S genes are being eliminated when homozygotes die of either malaria or anemia, but the heterozygotes survive better, and keep both kinds of genes in the population. When two heterozygotes (AS) have children, their genes combine into any of three combinations, AA, AS, and SS. So the weaker genotypes are continually born, only to die of one or the other disease.

This is an example that every biology student knows, but its lessons are seldom pointed out. First, this is natural selection in action: some genotypes survive and reproduce better than others. Second, this particular form of natural selection keeps both kinds of genes in the population, so that the species is genetically variable. Thus the existence of genetic variation in this population is not a hedge against the possibility that the environment may change. It is not an anticipation of the species' future needs. It is simply a mechanical consequence of the accidental fact that heterozygotes survive better. Third, the population may be said to have adapted to a malarial environment by having the S gene in high frequency, but at what a cruel cost! The price of keeping a malaria-adapted gene in the population is the death of anemic homozygotes, generation after generation. This sort of situation, repeated over and over in other adaptations, reveals something important about natural selection. It is an uncaring mechanical process, whereby the species

achieves adaptation at the expense of the death (or inadequate reproduction) of less well adapted organisms.

Henry Morris, the head of the Institute for Creation Research, has
written that "if evolution is true, then there have been three billion years
of suffering and death in the world of living things leading up to man.
. . . It does seem that, if God used evolution, He used the most wasteful,
most cruel process, that he could possibly conceive by which to produce
man. But the God of the Bible is not that kind of God!"[7] Indeed.
However, the sickle-cell case shows exactly such a "cruel process" in
action, but the impersonal process of natural selection, unlike a beneficent Creator, does not act for the good or comfort of the individual.

Indeed, if the world's species had been created by the wise and
loving deity imagined by the creationists, why should more than 99
percent of them be extinct? As Henry Morris says, "Billions of animals
suffered and died, for no apparent reason. Multitudes even of species
have died out."[8] But the fact is that natural selection is not omnipotent,
and does not provide insurance against extinction. It can even hasten the
demise of a species. The extraordinary case of the t gene in mice is a
case in point.[9] Male mice that are homozygous for a certain gene, t, are
sterile. But heterozygous male mice, which have one T gene and one
t gene, are fertile. Now we would expect, under the rules of genetics,
that half the sperm cells of such a mouse would carry the T gene and
half would carry t. But in fact the t form of the gene has some kind of
advantage, so that it is carried by about 95 percent of the sperm cells.
The t gene can therefore flood the population, and if the mouse population is small, it is possible for all the males that are born to be sterile
tt homozygotes. Such a population will become extinct, and apparently
this often happens. This is an instance of natural selection—one kind
of gene reproduces better than another kind—which results not in
adaptation but in the reverse.

It would be a mistake, then, to suppose that natural selection always
adapts a species to its environment. In fact, natural selection often has
nothing to do with adaptation. One form of natural selection is what
Darwin called sexual selection: evolution that is based purely on superior reproduction. Male fishes, birds, and mammals, for example, often
have extraordinary colors, crests, horns, and courtship behavior, all of
which make them not less but more susceptible to predators. However,
the more exaggerated the male's characteristics are, the more attractive
he is to females, so the genes for such characteristics spread in the
population. The reality of sexual selection has been confirmed by many
studies. For example, Leslie Johnson, who studies animal behavior,

showed that in a Central American species of beetle, males with unusually long snouts are greatly preferred by females.[10]

The peacock's train of feathers is a perfectly natural result of a process in which genes that affect his plumage either succeed or not, depending on the whim of the female's sexual preference—a process that doesn't in any way enhance the peacock's adaptation to anything except the act of reproduction. Do the creation scientists really suppose their Creator saw fit to create a bird that couldn't reproduce without six feet of bulky feathers that make it easy prey for leopards?

What we see exemplified by the *t* gene and the peacock's tail is this: whenever a gene, for whatever reason, can bequeath more copies of itself to subsequent generations than any other gene, it takes over—without forethought and without regard for whether it will be good for the individuals who inherit it or good for the species as a whole. Perpetuation of the species is not a cause but a result of reproduction. When this purely mechanical aspect of natural selection is fully understood, all sorts of puzzling phenomena fall into place, as they never could in a creationist theory. For example, if an environment remains fairly constant for a long time, a species may become highly specialized and dependent upon some particular food or habitat. When the environment changes, the species may be bereft of sufficient genetic variation to adapt to the change, with extinction the result. The ivory-billed woodpecker is now almost certainly extinct because it could not foresee that the dead trees in which it nested would be eliminated as forests were cleared. Selection of genes, if it causes adaptation at all, adapts a species to current environmental conditions, not future ones—and so most species that have ever lived are extinct.

Consider, further, the question of why plants and animals, including people, are mortal. Why aren't organisms capable of living forever? A Creator could presumably make a perfect organism that could live forever. But genes cannot. An organism can acquire only so much energy; the more energy it devotes to reproduction, the less it has for its own maintenance and survival. Thus organisms that reproduce early in life "wear out," and are apt to die. It turns out, if you work out the mathematics, that genes that promote reproduction will make more copies of themselves than those which merely prolong life. If this hypothesis is true, we should find that when a population is forced to evolve longer life span, the rate of reproduction will decrease. This is exactly what Michael Rose and Brian Charlesworth reported recently in *Genetics*. When they bred from flies that laid exceptionally great numbers of eggs late in life, the flies' descendants lived longer but reproduced less while they were young, and the overall rate of population growth declined.[11] Therefore, the struggle to reproduce more seems to be responsible for the evolution of senescence and death.

Thus evolutionary biology can account for senescence. Can a creationist theory do so? Would a beneficent Creator create the mayfly, in which the larva lives for years in a stream and then emerges with intricately constructed wings and legs only to die within half an hour, the time it needs to mate and lay eggs? We can explain by natural selection why an elm tree should produce millions of seeds each year, almost all of which die: for if the genes are competing for those few places where a seed could grow successfully, those genes will have the highest chance of success which flood the environment with copies of themselves, in a lottery that is stacked against them with odds of millions to one. Why should a Creator design such waste into His system?

Creationists might answer (though I'm not sure they've addressed the question) that it's all wisely designed. Elm trees make millions of seeds to feed birds and rodents. Animals die at a predetermined age to make room for the next generation. It is all part of the harmony of nature, the harmony that reflects the Creator's mind.

If this were true, we would expect to see harmony in nature, not struggle; indeed, we would expect to see animals sacrificing themselves for the good of their species, and even making sacrifices for the good of the natural community in which they live. If the theory of natural selection is true, though, organisms should have adaptations that serve purely for the survival and reproduction of the individuals who bear them, not for the good of any other individual or species. Darwin laid down the challenge in *The Origin of Species:* "If it could be proved that any part of the structure of any one species had been formed for the exclusive good of another species, it would annihilate my theory, for such could not have been produced through natural selection."

How has Darwin's challenge fared? No one has ever found a case of a species altruistically serving another, without any gain for itself. Consider those relationships between species, such as pollination, in which some altruism seems at first glance to be involved. Flowers produce nectar to induce animals to help them reproduce. Plants that don't need animals, such as wind-pollinated pine trees and grasses, don't produce nectar. In fact, some plants deceive animals, and save themselves the energy that goes into making nectar. Many kinds of orchid flowers, for example, are shaped and colored so as to look like flies or bees. Pollination occurs when a male fly or bee "thinks" it sees a female and comes in to copulate—with a flower.

What about cases of cooperative behavior within a species, in which animals help each other? Doesn't this provide evidence for selfless altruism? In every case that has been examined, the apparent altruism turns out to be a way in which "the selfish gene," as Richard Dawkins has described it,[12] promotes its own survival and reproduction, but it does so indirectly. The idea is that because close relatives inherit the same genes from their immediate ancestors, a gene that influences an individual to help its relatives can therefore improve the survival of the other copies of that same gene which the relatives carry. Suppose, for example, that a gene in an early mammal caused the female to produce milk. (This is a hypothetical example, because such genes haven't been identified. You can't identify a gene for a trait unless some individuals have the trait and others don't. But all female mammals have this trait. Males don't count, because they must have the same genes without expressing them.) Milk production wouldn't improve the mother's survival. But if young mammals who are fed milk survive better than those that aren't, the gene for milk production will increase in the population, because the offspring who received the benefit of milk will carry their mother's genes for this trait. The gene has an advantage because it is, in a sense, taking care of other copies of itself. This kind of natural selection, called "kin selection," will work only if relatives tend to stay close together, so that the beneficiaries of this seeming altruism are more likely to be relatives than nonrelatives. Of course, the animal doesn't "know" it's helping relatives instead of nonrelatives. It's just that genes for cooperative behavior increase only in those species that disperse so little that relatives tend to stay together.

Kin selection may account for the most thorough selflessness known in the animal kingdom. Most honeybees are sterile workers, slaving away for the good of the hive. They even have a barbed sting, so that they are eviscerated when they sting a large enemy and cannot withdraw without ripping themselves open.

Sterility and suicide: how could such characteristics evolve? By kin selection. Because of a peculiar pattern of inheritance in the bees and their relatives, a female bee shares more genes with her sisters than with her daughters. By helping her daughters survive, she would ensure the propagation of her own genes, just as a mammal does by providing milk. But if a bee helps her sisters (the future queens) to reproduce, she can help to propagate more copies of her genes than if she were to have daughters, because her sisters carry more copies of her own genes. Therefore, genes that promote defense of her sisters, instead of reproduction, can increase faster in a bee population than genes that promote reproduction alone. The validity of this theory is shown by the male bees, the drones. They share more genes with their own offspring than with their sisters, so it doesn't pay off, genetically speaking, to help their sisters. Drones, therefore, are purely selfish: they don't do any work for the colony and have only one object in life—to mate.

Kin selection, therefore, is one of the reasons why apparently unselfish behavior can evolve. When cooperative behavior is examined in detail, it becomes evident that it has evolved under conditions that would put animals into contact with their relatives, so that an animal doesn't dispense help to the species as a whole, but primarily to individuals that share its genes. Cooperative behavior is most typical of animals like wolves or monkeys that stay together in groups of relatives. It is typically *not* found in grasshoppers or flies or other species that move around so much that relatives are dispersed. In species that do have cooperative groups of unrelated individuals, the cooperation appears to stem from individual self-interest, as it often does among people who band together for a task, such as fighting a common enemy. Schools of fish, for example, exemplify what William Hamilton of the University of Michigan[13] has called "the selfish herd." The fish form a tight school when danger threatens, because each fish tries to get into the center of the school and put other fish between it and the enemy. By being a member of the school, each fish makes it more likely that someone else is going to fall victim. If species had been created to serve each other and build harmony in nature, we ought to see animals offering themselves to feed their predators. But nowhere in nature do we see self-sacrifice, except, as in the honeybee, to protect relatives.

The result of natural selection is usually adaptation. As we have seen, the process of adaptation does not necessarily serve the good of the species as a whole. Rather, it consists of the gradual increase of individuals who are better able to propagate their genes, often at the expense of less well endowed individuals. The result, however, is often an intricacy of organization that bears the appearance of design. If we visit industrial regions of England or the United States, we will find at the present time many species of black moths that sit on dark tree trunks during the day and are difficult to see. If we didn't know their history, we would be impressed by how beautifully they are designed to blend into their environment and escape detection by predators. But in this instance we know that in the nineteenth century these tree trunks were covered with light gray lichens, and that almost all moths were light gray at that time. As air pollution increased during the Industrial Revolution, the lichens died, and black mutants that had been rare in the moth populations increased, replacing the gray ones. In the 1950s, H. B. D. Kettlewell showed that the increase was due to differential predation by birds, which pick gray moths off dark trees far more readily than black ones.[14] The appearance of design is an illusion. The match between organism and environment is the outcome of a historical process of evolutionary change.

Adaptation, then, is "design" that has been wrought by the impersonal forces of reproduction, survival, and a changing environment. In fact, what the appearance of design indicates is invariably that adaptation has taken place. For example, we know from principles of physics that heat radiates from an object in proportion to its surface area. In hot regions, therefore, we expect mammals to have a large surface area in relation to their body size to dissipate excess heat; in cold areas, they should have a correspondingly low surface area. We find, accordingly, that jackrabbits in hot deserts have large ears and long legs, while Arctic rabbits have short ears and legs, so that their surface area is greatly reduced. We can interpret these features as adaptations to temperature, because they conform to the design that an engineer might have used to construct a rabbit with the right body temperature.

Conformity to optimal design is not, however, evidence for evolution. Quite the opposite; as the creationists incessantly argue, optimal design is exactly what we would expect of an intelligent designer. However, if natural selection were the cause of the appearance of design, we should instead expect animals and plants not to conform to engineering principles in any optimal way. Natural selection cannot invent the best possible genetic variations: it can only replace inferior genes with the best genetic variations that happen to be available when the environment poses its challenge. Thus we should expect to see many different genetic solutions to any adaptive problem.

One example is the resistance of insects to insecticides. Different populations of the same species of fly have adapted to DDT by using single recessive genes, single dominant genes, or multiple genes; and the physiological mechanism by which the genes confer resistance varies: some produce more of an enzyme that breaks down DDT, while others slow down the penetration of DDT into the body or into the nerve cells.[15] This principle explains why different species have different mechanisms of adapting to the same environmental factor. The chicks of sandpipers, for example, have complex color patterns that enable them to blend into the background, but every species of sandpiper has slightly different markings. There is no reason to think that any one color pattern is best: they are all probably equally good protection against predators. Similarly, there is no reason to suppose that African rhinoceroses are better off with two horns and Indian rhinoceroses with one: fighting and defense merely require horns, and the exact number probably isn't important. What could have possessed the Creator of the creation scientists to bestow two horns on the African rhinoceroses and only one on the Indian species?

Not only do species manifest a bewildering array of equally good adaptations, they are often quite poorly adapted. Lemmings and locusts have no adaptations to prevent overpopulation. The populations grow, exhaust their food supplies, and migrate by the millions, to perish in oceans or deserts. The potato leafhopper migrates by the millions every year from the southern United States to the north, and then perishes during the winter, to which it is not adapted. Look at the structure and physiology of species and you will find many instances in which species are not yet optimally adapted to their way of life, and other cases in which they are adapted to a former way of life that they no longer practice. The marine iguana of the Galápagos Islands spends much of its life diving beneath the waves for seaweed, but it has virtually no physiological or structural adaptations for life in water: it is essentially a terrestrial lizard, distinguished from the Galápagos land iguana merely by its slightly flattened tail. It can't hold its breath under water any longer than the land iguana can.[16] Conversely, look at a common dandelion and you will see a species adapted to its past. Most of the species of dandelions reproduce sexually, and have nectar and bright yellow petals that attract insects for cross-pollination. But the particular species of dandelion that grows in everyone's lawn is an anachronism: it reproduces entirely asexually, and does not need to be pollinated. Yet it still has nectar and yellow petals to which insects come, though they serve no function. They are useless characteristics, left over from the dandelion's sexual past.

In my own research I have found an equally intriguing situation. For several years, I and my associates[17] have been studying the fall cankerworm, a moth whose larva is destructive both to shade trees and to fruit trees. When the moths emerge in the fall, the females climb trees and release a sexual scent that attracts males for mating. Some of the females incorporate the sperm into their eggs, so that their sons and daughters are sexually produced; but the majority of the females do not use the sperm. Their offspring are all genetically identical to their mothers—which means, among other things, that they are all daughters. However, such females will not lay eggs unless they mate. Probably, as in some other species that have a similar system of reproducing, the sperm serve only to stimulate egg laying. Thus, these females are "sexual parasites." They use the males for their own ends, while the male gets nothing out of mating, since he does not get to pass on his genes. Clearly it would be advantageous for the males to reject these parasitic females, but they apparently haven't developed the ability to distinguish them. It would also be advantageous for the females to emancipate themselves and be able to lay eggs without mating, but they too haven't evolved this ability.

There is a further consequence of this process. Because the "parasitic" females have only daughters, the number of females in the population grows much more rapidly than the number of males, so that there are almost certainly not enough males to serve all the females. We have found in several parts of Long Island, New York, that there is only about one male for every hundred or so females. What this means is that the population can self-destruct. It will get to the point where there are almost no males, and then none of the females will be able to lay eggs. Perhaps for this reason the fall cankerworm goes through enormous population cycles, building up to great numbers and then crashing almost to extinction. What intelligent Creator, aiming for harmony and perfection, would have designed such a creature? Only in the context of natural selection does the fall cankerworm make sense. Because the "parasitic" females produce only daughters, each of whom can make more "parasitic" offspring, a gene for "parasitic" reproduction can increase in the population twice as fast as a gene that programs a moth to make both sons and daughters—and so the "parasitic" mode of reproduction takes over, even if it ultimately causes the demise of the population.

The creationist "argument from design" holds that adaptations are evidence of an intelligent designer. Thus, speaking of the similar adaptations of different organisms, the authors of *Scientific Creationism* write, "The evolutionist has to assume all such characteristics have developed by chance mutations and natural selection. Creationists explain them as structures designed by the Creator for specific purposes, so that when similar purposes were involved, similar structures were created."[18] But by the same argument, one might suppose that if organisms are not ideally adapted, if they have characteristics that are not adaptations, they could not have been intelligently designed—or at least the designer couldn't come up with the right materials or the right plan. A designer wouldn't equip organisms with useless appurtenances; yet every species has vestigial structures that may once have been adaptive but are adaptive no longer. Every species also has characteristics that are not now and never were adaptive—characteristics that are the "side effects" of genes that serve some other adaptive function.

We know from the study of genetics that genes produce biochemical substances that have many different effects on growth and development. Thus genes that promote growth of one part of the body usually affect the growth rate of other parts; hormones that produce an effect in the male also affect the female. As a result, genetic changes that occur for one adaptive reason usually have nonadaptive side effects. For example, when flies are selected for a change in the number of bristles, there is often a concomitant change in the shape of the female's reproductive tract, because some of the genes that affect bristle development also affect the reproductive organs. Therefore not every evolutionary change is an adaptation. Men's nipples serve no adaptive function, but they are easily understood as by-products of the same genes that cause the formation of women's breasts.

Similarly, an evolutionary change in body size can automatically cause various characteristics to become either more or less pronounced. Stephen Jay Gould has recently devoted a large book, *Ontogeny and Phylogeny*, [19] to this aspect of evolution. He describes, for example, how many species of very small clams retain the juvenile form of their larger relatives. The spines and ridges on the shell that develop as the larger species mature never develop in the small species. Conversely, when a species evolves to be gigantic, it can become almost grotesque, as various parts of the body, growing at different rates, become disproportionately large or small. The front legs of the carnivorous dinosaurs that walked on their hind legs show a general trend toward reduced size in larger animals. This trend was carried to extremes in *Tyrannosaurus rex*, the largest carnivorous dinosaur, in which the front legs were so tiny that they must have been almost useless. None of these characteristics make adaptive sense, or show evidence of design. They are almost surely genetic side effects.

The world of biology shows that animals and plants are not optimally designed; that some of their characteristics are mere physiological by-products of growth and biochemistry; that the path to adaptation is set by historical accidents of genetic variation; that struggle, not harmony, is rampant in the world; that the end to which almost every species comes is oblivion. Can we find the workings of the creationists' designer in these "designs"?

The most scathing attack on the argument from design was written long ago—not by an evolutionary biologist, but by Voltaire.[20] A quarter of the population of Lisbon, 30,000 people, were killed on All Saints' Day in 1755, when an earthquake toppled churches that were crowded with pious worshippers. Divine retribution for human sin, said the clergy, and Voltaire was outraged:

> . . . all sentient things, born by the same stern law,
> Suffer like me, and like me also die.
> The vulture fastens on his timid prey,
> And stabs with bloody beak the quivering limbs. . . .
>
> The man, prone in the dust of battlefields,
> Mingling his blood with dying fellow man,
> Becomes in turn the food of ravenous birds.
> Thus the whole world in every member groans,
> All born for torment and for mutual death.
> And o'er this ghastly chaos you would say
> The ills of each make up the good of all!
> What blessedness!

Responding to Rousseau's reply, that man himself must bear the blame for the tragedy of Lisbon, for man was meant for pastoral, not urban life, Voltaire sharpened his ridicule, in his most famous work, *Candide*. There Dr. Pangloss expounded on "this best of all possible worlds": "the nose has been formed to bear spectacles . . . legs were visibly designed for stockings . . . stones were designed to construct castles." There is a special providence in every atrocity that man and nature can visit upon us. If we admit the complexity and order of nature as evidence of design, then the designer must be charged with incompetence or malice. Either God can prevent evil but will not, or he would prevent evil but cannot.

Had Voltaire been a biologist, what an essay he could have written! Instead of "a special providence in the fall of a sparrow," he would have found animals slaughtered by the million by the ravages of inclement weather, disease, predation, and competition for limited food and hiding places. As Darwin put it, "What a book a devil's chaplain might write on the clumsy, wasteful, blundering, low, and horribly cruel works of nature!" How blasphemous it is to conceive of an all-good and omnipotent God, and to charge him with clumsy design and the extinction of millions of species. It is no wonder that the Manicheans conceived of the world as a battleground between a creator of good and a creator of evil.*

In the world of nature, there is neither good nor evil. The extinction of a comet astronomers recently sighted plunging into the sun is not a cosmic tragedy, it is just an event produced by mindless physical forces. Neither is the extinction of the pterodactyl tragic, nor is the struggle for existence that causes evolution either good or bad. It just is. Species arise throughout the ages, "but time and chance happeneth to them all." So what is chance, and what role does it play in evolution?

---

* I am well aware that religious explanations for imperfections in the creation attribute them to malign spirits such as devils, or to Adam's fall. "Scientific creationists" have not explicitly invoked these explanations, so I have not addressed them here except in an endnote.[21]

SEVEN

CHANCE
AND
MUTATION

The rise of creationism is politics, pure and simple; it represents one issue (and by no means the major concern) of the resurgent evangelical right. Arguments that seemed kooky just a decade ago have re-entered the mainstream.
—STEPHEN JAY GOULD, 1981*

As Tom Stoppard's play *Rosencrantz and Guildenstern Are Dead* opens, the title characters are flipping their sixty-ninth coin, and for the sixty-ninth time it comes up heads. They continue. The ninety-second coin comes up heads. They have played this game before, and they almost always get heads. Guildenstern sees the problem: the law of probability has been violated. "The equanimity of your average tosser of coins depends upon a law, or rather a tendency, or let us say a probability, or at any rate a mathematically calculable chance, which ensures that he will not upset himself by losing too much nor upset his opponent by winning too often. This made for a kind of harmony and

---

* Stephen Jay Gould, who teaches paleontology and the history of science at Harvard University, has been active in the defense of evolution against creationism.

a kind of confidence. It related the fortuitous and the ordained into a reassuring union which we recognized as nature." Guildenstern expects chance to be a part of nature, and he recognizes that chance should result in predictable patterns. About half the coins should have come up tails. As the play proceeds, a series of events follows. By "accident," Hamlet kills Polonius. Claudius sends him under guard by Rosencrantz and Guildenstern to England, where he is to be murdered. Hamlet changes the letter, charging the English king to execute his guardians. By chance, pirates attack and Hamlet escapes. Rosencrantz and Guildenstern sail inexorably toward death. "There must have been a moment, at the beginning, where we could have said—no. But somehow we missed it." Was their death determined from the beginning, or was it chance?

The meaning of chance, or randomness, is a problem that has preoccupied philosophers and mathematicians for centuries. While there is no need to delve into the intricacies of what chance is, we can get close enough to the issue to be able to use the word in a reasonably meaningful way. First, "chance" is not a simple antonym to "purpose," although this is strongly implied in a lot of creationist writing. In *Scientific Creationism*, for example, we are told that evolution "may tend to rob life of meaning and purpose in view of the implanted concept that the student is merely a chance product of a meaningless, random process."[1] Here and in other creationist writings, if something happens without purpose or design, it is seen as the result of chance. This is not, however, the way "chance" is used either in science or in everyday life.

There is no purpose in the movement of the tides, or in the pattern of the ripples in a stream, but we do not attribute them to chance. We attribute them to physical causes—the gravitational attraction of the moon, or the distribution of rocks in the streambed. Neither does chance mean "without cause." When we flip a coin, we trust that physical causes determine how it falls. It is only that we are ignorant of the exact physical forces operating on it, and so cannot predict how it will land. Before a sperm and egg united to make me, there was presumably a fifty–fifty chance that the fertilized egg would be female; but a chain of physical events determined that one particular sperm, bearing a Y chromosome, would be the "lucky" one. In general, then, we say that chance operates when physical causes can result in any of several outcomes, but we do not have sufficient knowledge to predict what that outcome will be in any particular case. At least some of what we call chance, then, is a name for our ignorance. It is still a serious philosophical problem whether it is possible to have enough knowledge for perfect prediction —in which case chance would vanish.

Although we more or less equate "chance" with "unpredictability," like Rosencrantz and Guildenstern we also assume that it is possible in our daily lives to make powerful predictions about how a system will behave *on average*, even when we can't predict any single event. Scientists make the same assumption. The laws of genetics, for example, are among the most useful predictive generalizations in biology. We cannot predict whether any particular baby will be male or female, blue- or brown-eyed, but we can say that about half of any large number of children will be girls, and that about a quarter of the children born to heterozygous brown-eyed parents will have blue eyes.

Sometimes when we obtain additional information we can make better predictions; that is, events will seem less random. Knowing some of the causes of an event is a good way of improving prediction. Thus if asbestos is known to be a cause of skin cancer, our ability to predict whether someone will get skin cancer or not is greatly enhanced by knowing whether or not the person works with asbestos. But we may find that asbestos workers contract cancer with equal frequency whether they are men or women. We would say then that sex is not a "cause" of susceptibility to skin cancer. We would also say that skin cancer strikes at random with respect to sex, but not at random with respect to occupation. This distinction will be of some importance when I discuss the randomness of mutations.

It is pretty obvious that both chance (unpredictable) and nonrandom (predictable, or consistent) factors operate at the same time: any of us may suffer an automobile accident from the unpredictable behavior of other drivers, but we are much more likely to crash if we habitually drive after drinking. Thus the chance that an event will happen is altered by all sorts of factors. This implies that if someone builds an argument on the basis of probabilities, the assumptions of the argument have to be watched very carefully. For example, creationists claim that the probability of life evolving from nonlife is vanishingly small. One of their arguments is that spontaneously formed nucleotides would be so dilute in the primitive ocean that they would have hardly any chance of aggregating into nucleic acids. But this ignores the fact that chemicals will accumulate in some places even if in the ocean as a whole they are greatly dispersed; or that organic compounds commonly adhere to surfaces, and so would be concentrated on the surfaces of sand grains or clay particles.

For a more subtle example, consider the argument[2] that the first

DNA molecule could not have evolved by chance. Since each site on the molecule can be occupied by one of four nucleotide bases, the chance that a particular DNA molecule 1,000 nucleotides long would be formed is only 1 out of $4^{1,000}$, or about 1 out of 10 followed by 600 zeroes. Quite so, if the first DNA molecule had to have any particular sequence of nucleotides. But *any* sequence of nucleotides would replicate itself, as the experiments by Manfred Eigen that I described in Chapter 4 show. Any of a large number of mutational changes in the initial sequence could improve the stability of the molecule (its "survival") or improve its rate of replication, and cause subsequent evolution of the population of molecules. Any particular mutation that enhances the survival or the replication rate may occur with a very low probability, but the chance is much higher that one or another of the many possible mutations that have this effect will occur. Thus a "new, improved" DNA molecule may be quite likely to evolve, but the one that actually evolves will be only one of the many improved DNA molecules that could have occurred.

If we had been looking at mammal-like reptiles in the Triassic, we might have supposed it almost infinitely unlikely that some particular series of genetic events would occur that would result in the elephant. But the elephant is the result of only one out of a virtual infinity of possible evolutionary sequences that could have occurred, most of which didn't—just as you are one out of millions of genetically different children your parents could have had, but didn't. This is *not* to say that all conceivable evolutionary paths from the mammal-like reptiles were equally likely. They were not, because many conceivable evolutionary changes would not be permitted by natural selection. An elephant with spindly legs would be an adaptive impossibility. But out of the many possible adaptive paths that the mammal-like reptiles could have followed, only a few actually became realized, and which ones were realized must have been strongly influenced by chance.

Almost all phenomena are affected both by chance and by more consistent, or "deterministic," factors. For example, gas molecules move randomly and cause the random movement (called Brownian movement) of small particles with which they collide. If you watch a dust particle in a beam of light, you will see it move both up and down by Brownian movement. But the dust particle is also subject to a deterministic force, gravity, that pulls it down. The force of gravity on a dust particle is small compared to the random impact of gas molecules; but as the weight of a particle increases, random effects, although still present, become less important compared to the deterministic action of gravity. Similarly, both deterministic factors, especially natural selection, and random factors act to cause evolution.

By far the most important way in which chance influences evolution is in the process of mutation. Mutation is, ultimately, the source of new genetic variations, and without genetic variation there cannot be genetic change. Mutation is therefore necessary for evolution. But it is not sufficient. A new mutation exists at first in just one individual of the species, and then in that individual's offspring. Therefore it is carried by just a few individuals at any time, unless something makes those individuals reproduce more than others, so that the mutation becomes more common. That "something" is either genetic drift or natural selection. Evolutionary change requires at least two ingredients—mutation and either genetic drift or natural selection. Creationists often accuse biologists of attributing all of evolution to chance, and this would be true if mutation were the whole story. But natural selection is part of the story, and it is not chance. Quite the opposite, it is this factor that shapes order out of mutational chaos.

At the molecular level, there are many kinds of mutations. Perhaps the most common type is a change in one or more of the nucleotides that make up a gene, which is a string of thousands of pairs of nucleotides in a ladderlike arrangement. The exact sequence of four kinds of nucleotides determines the exact sequence of twenty kinds of amino acids in a protein that is produced by the DNA. The amino acid sequence of the protein determines its biochemical function, which in turn affects the development, form, and physiology of the plant or animal. Some mutations can therefore have profound effects. They can alter the structure of a critical protein so much that the organism becomes severely distorted and may not survive. Other mutations may cause changes in the protein that don't affect its function at all. Such mutations are adaptively neutral—they are neither better nor worse than the original form of the gene. Still other mutations are decidedly advantageous. Between these extremes, there is a complete spectrum of effects, and this is probably where most mutations fall.

Many mutations in fruit flies, plants, and other organisms are known that increase or decrease the activity of enzymes and proteins to a greater or lesser degree. At the level of the whole organism, most mutations have slight effects. In fruit flies, for example, most mutations cause slight increases or decreases in the rate of growth, body size, the length of wings or legs, the number of bristles, the ability to detoxify DDT and other poisons, and so on. Some mutations, to be sure, have drastic effects, like the *vestigial* fruit fly gene that reduces the wings to little nubbins. Because such mutations are easy to measure, they have been studied extensively; but the vast majority of mutations have much more subtle effects. It is possible, as Alan Robertson of Edinburgh, Terumi Mukai of Wisconsin, and many others have done, to make a population of fruit flies that is genetically completely homogeneous, and to measure new genetic variation in the population as it comes into existence. If you look for the appearance of very deleterious or lethal mutations, you will find them; but if, as Robertson has done, you look for new variation in the number of bristles or any other feature, you will find that as well. Such variation increases from generation to generation, as subtle new mutations come into existence.[3]

Most biologists believe that mutations occur simply because organisms can't prevent them. From generation to generation, DNA molecules replicate themselves by separating the two sides of the DNA "ladder"; enzymes then insert each of the four kinds of nucleotides into the right positions, so that the two identical ladders are formed from the original. Sometimes the wrong nucleotide is inserted into a particular position. There are "repair enzymes" that help to correct such mistakes, but they don't correct them all. Such mistakes presumably occur because chemical reactions don't happen with perfect precision every time. In fact, a variety of chemicals such as caffeine can increase the mutation rate, as can various forms of radiation.

If mutations, then, have physical causes, in what sense are they random? Only in the sense that the adaptive "needs" of the species do not increase the likelihood that an adaptive mutation will occur; mutations are not directed toward the adaptive needs of the moment. As the geneticist Theodosius Dobzhansky said, "Only a vitalist Pangloss could imagine that the genes know how and when it is good for them to mutate."[4] This has been shown elegantly in experiments with bacteria and fruit flies. For example, Joshua Lederberg did an experiment[5] in which he grew thousands of colonies of genetically identical bacteria from a single bacterial cell that was unable to survive in the presence of streptomycin. He divided each colony of cells in two, and grew one half with and one half without streptomycin. A few of the colonies survived on streptomycin, because they carried new mutations for streptomycin resistance. When the unexposed halves were tested, the "sister" halves of the resistant colonies also proved to be resistant, even though they had never been exposed to streptomycin. Thus the mutations for streptomycin resistance occurred before exposure to the drug, and were inherited both by the exposed and unexposed halves of each colony that had a resistant mutant ancestor. If the bacteria encounter streptomycin, the mutation is clearly adaptive; if they don't, it isn't. Mutations have causes, but the species' need to adapt isn't one of them.

Many experiments other than Lederberg's clearly prove that there are adaptive mutations. Patricia Clarke selected mutations of *Pseudomonas* bacteria that enabled them to grow by metabolizing novel organic chemicals.[6] Francisco Ayala found that the size of a dense population of fruit flies increased as genetic changes occurred that enabled the flies to use their food more efficiently. Populations that were irradiated grew to even higher size; the irradiation increased the amount of genetic variability that the populations could use for adapting to their food.[7] Paul Hansche observed a mutation arise in yeast that doubled the gene that codes for the enzyme acid monophosphatase. The mutant yeast had an increased amount of the enzyme, which enabled the mutant to obtain more organic phosphate from its environment and gave the mutant an adaptive advantage over the nonmutant yeast.[8]

One of the most extraordinary claims made by creationists is that adaptive mutations do not occur. In *Scientific Creationism*, for example, we read that "the phenomenon of a truly beneficial mutation, one which is *known* to be a mutation and not merely a latent characteristic already present in the genetic material but lacking previous opportunity for expression, and one which is permanently beneficial in the natural environment, has yet to be documented."[9] Now the only way we can know that a genetic variation is a new mutation is to see it arise in a genetically uniform population, which means that the demonstration that new mutations occur can only be achieved in the laboratory. But it is an empirical fact that natural populations of yeast and other microorganisms commonly do not have an abundant supply of phosphate for growth. Can anyone possibly doubt that Hansche's yeast mutation would have been as adaptive in nature as it was in the laboratory? Since exactly the same kinds of genetic variations that are seen to be adaptive in natural populations have repeatedly been observed to arise by mutation in the laboratory, is it plausible to imagine that all these variations were simply bestowed upon species by a Creator?

One of the many mistakes the creationists make is that they seem to think that a mutation must almost always be a bad thing. Thus Duane Gish claims that "the mutations we see occurring spontaneously in nature or that can be induced in the laboratory always prove to be harmful. It is doubtful that of all the mutations that have been seen to occur, a single one can definitely be said to have increased the viability of the affected plant or animal."[10]

In truth, whether a mutation is "good" or "bad" depends on the environment. In a species already well adapted to its environment, many mutations will take the species away from its optimum condition and thus prove harmful. The very same mutations could, however, become beneficial if the environment were to undergo a change. Several kinds of grasses, for example, have genes for lead tolerance that are disadvantageous in the absence of lead. Somehow they reduce the grasses' ability to compete with other plants. But in the vicinity of lead mines these genes provide a tremendous advantage, and lead-tolerant populations have evolved in such areas within the last thirty years.[11]

Because most mutations have only a slight effect on each of the characteristics that they alter, almost all characteristics display what is called continuous variation. There is a complete spectrum of height and hair color, for example, in many human populations. Typically there are dozens of genes that affect a characteristic, each existing in different forms that alter the characteristic very slightly. Suppose for the sake of example that three genes A, B, and C affect height. Every individual has two representatives of each gene. Let the average height of a plant with genotype *aabbcc* be 50 inches. Then if a capital letter represents a mutation that adds an inch, a plant that is *AaBbCc* will be 53 inches. Now suppose each such mutation has just come into existence, so that it is very rare in the population. When the genes are shuffled about during sexual reproduction, very few individuals will inherit more than one upper-case mutation, so most of the plants will be about 50 or 51 inches tall. It is possible for a 56-inch (AABBCC) plant to be produced, but this will be extremely uncommon as long as the upper-case genes are rare.

If, however, 51-inch plants, which carry either the mutation A or B or C, survive somewhat better than 50-inch plants, each of these mutations will have been distilled or concentrated in the population, so that they become more common. When these plants mate with each other, some of their offspring are now more likely to inherit two or more upper-case genes and be 52 inches or taller. The same process, if repeated again and again, would make the upper-case genes so common that a 56-inch offspring would become likely rather than unlikely. The population will then have evolved—and quite quickly, too—from an average height of 50 inches to an average of nearly 56, well beyond the range of variation in the original population.

I cannot stress too strongly that this hypothetical mode of inheritance—many genes, each adding or subtracting a little bit—is exactly the way most characteristics are inherited. The selection process I have described has enabled geneticists to alter the milk production of cows, the yield of corn, the resistance of crops to diseases, and almost every conceivable characteristic of the long-suffering fruit fly: wing length, mating behavior, sensitivity to temperature, rate of genetic recombination. It doesn't matter whether the selection is imposed by a geneticist who wants to breed taller corn, or by competition in nature that favors plants able to shade out their competitors; if the mutations are there, and if the selective pressure in favor of them is strong enough, the population will evolve.

How continually does such genetic variation arise by mutation? Rates of mutation are studied by scoring how often a particular mutant appears—how often white-eyed fruit flies arise from red-eyed ancestors, for example. Because a single gene is composed of thousands of nucleo-

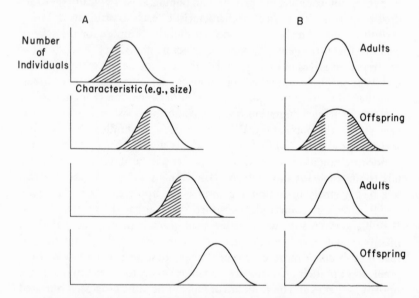

FIGURE 23. Two common forms of natural selection. In both, the cross-hatched area represents young individuals (offspring) that are less capable of survival, or adult individuals that are less capable of reproducing, than the portions of the curve that are clear. The series (A) at left represents directional selection, in which the superior reproduction and survival of larger individuals causes a shift in the characteristic toward larger size in the subsequent generation. The series (B) at right represents stabilizing selection, in which the average size remains the same from one generation to the next, because average individuals have the highest chance of survival.

tides, this eye-color gene could experience thousands of different kinds of alterations, not all of which would change the pigmentation enough for us to notice. Therefore our measures of mutation rate tend to be underestimates. Measured in this way, however, an average gene seems to mutate at a rate of about 10^{-5} per generation. That is, about one in every ten thousand sperm cells or egg cells produced in every generation during reproduction will carry a new mutation of a particular gene.

That doesn't sound like many mutations. But as I write these words, there are at least 200 gypsy moth caterpillars on every oak tree on campus, and I dare say at least 10 million on Long Island. Each caterpillar has two representatives of every gene, so if 1/10,000 of these genes mutate this year, there will be 2,000 new mutations of an average gene, just on Long Island. If gypsy moths have 10,000 different genes on their chromosomes (based on estimates from fruit flies), 2,000 × 10,000 = 20 million new mutations, affecting all kind of characteristics, will arise in these moths just this year, just on Long Island. Theodosius Dobzhansky and his coauthors, in their textbook on evolution,[12] estimate that the total number of mutations that have arisen in the human population of the world during the present generation is about 8×10^9: 8 billion. At this rate, a genetically homogeneous population would take only about 1,000 generations to achieve the same level of genetic variability that most species contain in nature. This isn't a hypothetical statement; it has been confirmed experimentally by Alan Robertson and his colleagues at the Institute of Animal Genetics in Edinburgh, in their work on fruit flies.[13]

So is it true, as the creationists claim, that good mutations are vanishingly rare? Certainly it is true that many, many mutations are harmful. But if even only one hundredth of 1 percent of all mutations are beneficial, 20,000 of them should crop up in the gypsy moths of Long Island just this year. Moreover, the early geneticists significantly underestimated the proportion of beneficial mutations, because they studied a biased sample of possible mutations, the ones that drastically change an organism instead of modifying it only slightly.

At least some creationists have such a profound misunderstanding of genetics that they make statements like this one:

> That the net effect of mutations is harmful, rather than beneficial, to the supposed progress of evolution, is made transparently clear by the zeal with which evolutionists for decades have been trying to get mutation-producing radiations removed from the environment! . . . It does seem that, if evolutionists really believed that

evolution is due to mutations, they would favor all measures which could increase the rate of mutations and thus facilitate further evolution.[14]

There are so many errors in this statement it is hard to know where to begin. Let us, for the sake of argument, grant the premise of the argument, that a majority of mutations are harmful, even though this probably isn't true. If 60 percent of the mutations are harmful, and every newborn animal has one new mutation, we could suppose that 60 percent of them die because of their new mutations. But in the vast majority of species, far more than 60 percent of the offspring die anyway, because of predators, weather, shortage of food, and so on. Individuals with deleterious mutations are likely, almost by definition, to be weaker and hence more susceptible to these factors. This "disposable" fraction of the population, then, carries with it the harmful mutations. The small fraction of the population that does survive will carry the beneficial mutations, and they will be propagated through the population by natural selection. Thus even if the "net effect of mutations is harmful," the net effect of mutation plus natural selection is beneficial.

So why don't evolutionists want to build leaky nuclear reactors all over the place? First of all, who wants to "facilitate further evolution"? This assumes that evolution is desirable. But evolution isn't either good or bad, any more than a volcanic eruption is. It just happens. We do make value judgments, though, about human fortunes. Should we wish to increase the human mutation rate? Obviously not. We don't consider 60 percent or even 1 percent of our children disposable, and we view every premature death and every genetically maimed child as a tragedy. So even if radiation could increase the chance of new mutations that made us resistant to cigarette smoke and air pollution, the harmful mutations that would occur would outweigh, in our minds, the beneficial ones.

There is one more fallacy in the above quotation that leads us into an important topic. The fallacy is that the rate of evolution would go up if the mutation rate were to increase. But just because mutation is an essential ingredient of evolution, it doesn't follow that the rate of mutation governs the rate of genetic change. You can't drive a car without engine oil, but the speed you drive doesn't depend on the amount of oil you put in. The fact is that most species could adapt a great deal to environmental changes for quite a long time even if the process of mutation could somehow be stopped right now. This is simply because a new mutation doesn't face only the two alternatives of instant elimination or instant takeover. Even a quite harmful mutation may hang around for a long time in a population before it is eliminated. In the meantime, the environment could change and make it beneficial. Even an advantageous mutation doesn't necessarily take over a population completely; it can increase and then for a variety of reasons become stabilized at an intermediate percentage of the population. In an African swallowtail butterfly, for example, different colored forms exist that are advantageous because they look like other, distasteful, species of butterflies, and so are avoided by birds. The percentages of the differently colored genotypes can shift rapidly if one or another of the other butterfly species becomes more common.

Therefore even if favorable mutations arise slowly, they accumulate in a population, so that a great deal of genetic variability builds up. Thus, almost every species ever examined has been found to contain an immense amount of genetic variability in various kinds of enzymes, physiological properties such as heat tolerance, behavioral characteristics such as mating behavior and feeding preferences, and structural properties such as the size and shape of almost every part of the body. Species therefore have the potential to be genetically altered to a very considerable extent when the environment imposes new pressures. It is hardly surprising, then, that many cases of rapid evolution have been observed. I have already mentioned the evolution of insecticide resistance in insects and lead tolerance in grasses. In the hundred years since the European house sparrow was first released in North America, it has diversified into different races: for example, small pale sparrows in the Southwest, and larger, darker ones in the North.[15] Within the last thirty years, the codling moth, which used to attack only apples, has evolved races adapted to plums and walnuts.[16]

All this genetic variation has been revealed only in the last twenty or thirty years, and it has greatly altered our view of how rapidly evolution occurs. Darwin supposed that adaptive changes would have to occur with excruciating slowness, but he was wrong. Most of the time a species may not be evolving very much, as long as it is reasonably well adapted to an environment that isn't progressively changing. However, when a substantial shift in the environment occurs, the already existing abundance of genetic variation sometimes permits a species to change very rapidly, in dozens or hundreds of generations. It is simply wrong to say, as the creationist Duane Gish does,[17] that slow and gradual change is an essential feature of the process of evolution. Gradual perhaps, but not necessarily slow.

Thus species contain abundant genetic variation that often enables evolution to be very rapid when the environment changes. But whether or not the species contains the "right" variations is very much a matter of chance. For every species of grass that has recently adapted to the toxic soils around lead or zinc mines, there are hundreds of species of plants that have not become adapted. Apparently, helpful mutations haven't occurred in them in the recent past. The altered environment simply can't evoke adaptive mutations; they occur by chance.

Chance plays many other roles in evolution. One of the most important takes place when a mutation occurs that is neither better nor worse than the original form of the gene. For example, some of the amino acids that make up a protein simply serve to fill space, so that the protein has the right shape. In many cases it doesn't seem to matter which amino acid occupies the space, as long as something is there. As a result, many mutations that alter the amino acid sequence of a protein neither reduce nor improve the protein's function; they are adaptively neutral, or at least they seem to be. It is possible, of course, that they alter the protein's function slightly; but if so, the effect is too slight to be detected by any methods that have been devised so far.

The percentage of the population that inherits such adaptively neutral mutations will increase or decrease purely by chance—the process of genetic drift that I described in Chapter 2. A neutral mutation of this kind can increase by chance in some populations or species and decrease in others, so that they eventually become genetically different—although adaptively equivalent. The rate at which these changes occur is faster in small populations than in large ones. Thus if the theory of genetic drift is true, small populations should be genetically more different from each other than large populations. This is just what Robert Selander, now at the University of Rochester, found when he examined the percentages of two forms of hemoglobin in mice.[18] Each barn in the area of Texas where he worked had a separate population of mice. Some populations were small, and they differed from one another quite considerably in terms of how many mice in a barn had one form of hemoglobin versus the other. But large, separated mouse populations all had about the same percentages.

Although there is still a great deal of controversy over the matter, many population geneticists believe that a good deal of the genetic variation in various proteins that has been found in many species is of this neutral kind, and that many of the differences in proteins among species have evolved by genetic drift. Thomas Jukes and Jack King, who were among the first to advance this theory, referred to it as "non-Darwinian evolution," since chance, rather than natural selection, seems

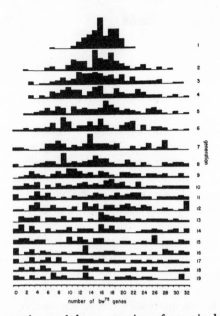

FIGURE 24. An experimental demonstration of genetic drift. 105 separate laboratory cultures (populations) of fruit flies were set up, each of which was initially genetically identical. Each population was started with 16 flies (8 females and 8 males). Each fly carries two representatives of a gene that affects eye color—either *Bw*, for red eyes, or *bw*, for brown eyes. So there were 32 gene copies represented in each population, of which half were initially *Bw*, and half *bw*, in each of the 105 populations. Thus each population had 16 *bw* genes initially. This series of graphs represents what happened over the course of 19 generations to the 105 populations. Each new generation was set up by 8 females and 8 males taken at random from the offspring of the previous generation. The topmost graph shows how many of the populations had any given number of *bw* genes after one generation: many of the populations still had 16 *bw* genes, but some had drifted at random so they had either more or fewer: one population had only 7 *bw* genes (and therefore, by subtraction, 25 *Bw* genes). By generation 2, the populations had spread out, so that they ranged from 2 to 29 *bw* genes. In subsequent generations the populations came to differ more and more in the percentage of *bw* genes, so that by generation 19 they were almost uniformly spread over the entire range. The percentages of the two kinds of genes had come to differ by chance from one population to another. If these changes in percentages had been caused by natural selection, most of the populations would have shifted consistently in one direction. (From P. Buri, *Evolution* 10:388, [1956].)

to be the major reason for the changes. But neither they nor anyone else claim that *all* evolution is non-Darwinian: adaptive features such as claws and wings clearly were developed by natural selection rather than genetic drift.

But just as gravity and Brownian movement may both affect the motion of an airborne particle, chance and natural selection often work simultaneously, and certain evolutionary phenomena can be understood only if we take both into account. Many populations of houseflies throughout the world have evolved a resistance to DDT—an adaptation that has come about by natural selection. In some populations, however, the adaptation is provided by a dominant gene; in some, by a recessive gene; in some, by a number of genes, each with a small effect. The physiological mechanism by which the genes act also varies: flies can be resistant, for example, either by having developed an enzyme that degrades DDT or by having altered the cell membranes so that DDT is less able to penetrate the tissues. These are alternative adaptive mechanisms. Which one developed in a particular population must have depended on which mutations happened to have been present in the population when it became exposed to DDT—and this is very much a matter of chance. Thus, chance initially determines what genetic variations will be acted on by natural selection to develop an adaptation.

When we extrapolate this principle of indeterminacy to long-term evolution, we can understand why different organisms have evolved different "solutions" to similar adaptive "problems." By chance, they had different genetic raw materials to work with. It is doubtless adaptive for male frogs to have a vocal sac that enables them to produce resonant calls that attract females. But whether a frog developed a single sac in the middle of the throat, as in the bullfrog, or a pair of sacs on either side, as in the leopard frog, may have been affected by what mutations first occurred by chance in the ancestor of each species.

If chance is a name for the unpredictable, then almost any historical event is affected by chance. Would Hamlet's mother, watching him stab Polonius through the arras, have predicted that this would be one in a chain of events leading to the death of Rosencrantz and Guildenstern? If you had been on the island of Mauritius in the mid-Tertiary, would you have predicted that the pigeons there would evolve into flightless dodos and then become extinct in the seventeenth century because they were easy prey for sailors? If you had seen a bipedal ape on the plains of Africa in the Pliocene, could you have predicted that this feature would prove crucial in the evolution of a larger brain and the development of human culture? Probably not; for in all such instances, the event that we recognize in hindsight as a "cause" might have been followed by other events leading to a different outcome. All of evolution, like all of history, seems to involve chance, in that very little of what has happened was determined from the beginning.

The mind that cannot abide uncertainty is troubled by the idea that the human species developed by "chance." But whether we evolved by chance or not depends on what the word means. We did not arise by a fortuitous aggregation of molecules, but rather by a nonrandom process—natural selection favoring some genes over others. But we are indeed a product of chance in that we were not predestined, from the beginning of the world, to come into existence. Like the extinction of the dodo, the death of Rosencrantz and Guildenstern, or the outbreak of World War I, we are a product of a history that might have been different.

EIGHT

THE ORIGIN
OF SPECIES

The creationists attempt to ride herd on the libraries, too, and do their best to pull out every book that doesn't suit them. And they want "equal time"? Don't kid yourself. They want all the time there is. One can see why, too. Their case is so weak, so nonexistent, in fact, that the only way they can feel sure of maintaining it is to make sure their victims *never hear of anything else.*

—Isaac Asimov, 1981*

One of the most influential metaphors in all of literature is Plato's famous myth of the cave, in *The Republic*. Imagine, he says, that some men have been chained for life in a cave, where they face only one wall. A succession of objects that they cannot see passes behind them. All they can see are the shadows cast on the cave wall by the light that enters the cave's mouth. "To them, the truth would be literally nothing but

---

* Isaac Asimov, a biochemist, is a prolific writer of books on science for the nonscientific public.

the shadows of the images." Like those men, says Plato, we take to be reality only the imperfect shadows of the realities, or ideal forms, that exist in the realm of spirit. Every circle a man can draw is an imperfect representation of the ideal circle we can conceive of, and which exists in the world of essences.

Arthur Lovejoy, in *The Great Chain of Being*, expresses Plato's notion this way: "The true objects of rational knowledge, the only genuine realities, are the immutable essences of things—of circles and all figures, of all bodies, of all living things, of all affections of the soul, of the good and the fair and the just. These essences are never identical with the fleeting objects which are their sensible manifestations."[1] Thus the doctrine of essentialism was born: the idea that there is a true unchanging essence to everything in the world, which determines its properties and abilities. Fires partake of the essence of fire, which is to burn; water of the essence of water, which is to flow. The object of science, then, is to discover the "true essence" of things. People may vary, but underlying this variation there must be a true essence of humanity, waiting to be defined and discovered. The object of classification is to see past the superficialities to the true essence of species and to arrange them in accord with their ideal representations, which only God can see.

The doctrine of essentialism is not very familiar to us now, but it was an integral part of Western thought for almost 2,000 years. It gave rise to the idea that the objects of nature consist of real "kinds," and that variation could not trespass beyond the borders of these kinds, each of which had some objective, ideal reality. As late as 1874 we read in John Stuart Mill's *A System of Logic* that "the universe, so far as known to us, is so constituted, that whatever is true in one case, is true in all cases of a certain description; the only difficulty is, to find the description. Kinds are Classes between which there is an impassible barrier."[2] Among these kinds were species (Latin for kind) of animals and plants.

In modern science, there is little room for essentialism. We recognize classifications to be arbitrary constructs of the human mind, not reflections of necessary divisions of reality. A "kind" of animal or plant is merely defined as such because of some particular set of characteristics, each of which can vary. If these vary enough, we will reclassify that animal or plant into another "kind." There is no immutable essence underlying the "superficial" characteristics; the "superficial" characteristics are all it has.

In the eighteenth century there was a considerable debate among the "monogenists" and the "polygenists" as to the nature of human races.[3] The monogenists held that human races were all the same "kind," and had somehow diverged from one original human race. The polygenists held that human races were different species, or kinds, that must then differ in their very essence. For them, the categories "Oriental," "white," and "Negro" were real, even if it should sometimes be difficult to classify an individual. For a modern biologist, the classification of humans into races is merely a convenient way to describe average differences in hair, skin color, and other physical characteristics. The variation is not discrete: each "race" that is recognized blends gradually into other races, and different characteristics vary in different ways. Thus, it is utterly arbitrary whether we recognize five human races or five hundred. It depends on how many characteristics are used for classification, and how finely the distinctions are made.

The notion of essentialism hasn't disappeared from popular thought, however. It persists as stereotypy and racism. The very essence of essentialism is the assumption that every member of a "kind" must share the features thought to define that kind. The supposition that blacks are lazy or that Jews are avaricious is a pernicious modern version of the centuries-old racism that fed on ancient essentialist ideas.

The notion that each species or race has an immutable, unvarying essence is completely dispelled by Darwinian evolution. For Darwin, there was no distinction between the superficial and the essential characteristics of humans or any other species; all characteristics are variable. The most essential feature of a species can be modified in time, and the most superficial or accidental variation can in time become typical of the group. But this notion is not acceptable to those twentieth-century essentialists who call themselves "scientific creationists." They believe in real essences. Thus the peppered moth, *Biston betularia*, has evolved in the last hundred years from gray to black, but to the creationist Duane Gish, this is not really evolution: "These moths today not only are still moths, but they are still peppered moths, Biston betularia,"[4] and no "real" evolutionary change occurred. Geneticists may develop enormously different breeds of chickens or corn, but according to Gish no real change has taken place.[5] He goes on: "By evolution, we do not mean limited variations that have taken place within a distinct, separate kind, and which have not led to the origin of a basically different form of life."[6] For Gish, genetic changes in each of the characteristics of a species aren't "real" evolution if they don't alter the species into a "distinct, separate kind."

What, then, are these "kinds"? Gish tells us that "a basic animal or plant kind would include all animals or plants which were truly derived from a single stock. In present-day terms, it would be said that they have shared a common gene pool."[7] There is, of course, a circularity in this definition. If biologists could show to Gish's satisfaction that lizards and snakes were derived from a single stock, they would then presumably qualify as a single "kind." Gish is aware of this difficulty: "We cannot always be sure, however, what constitutes a separate kind."[8] He is quite sure that monkeys, apes, and men are different basic kinds; but he allows that the various species of Galápagos finches, which are classified as different genera because they differ so greatly from each other, might all represent one kind, for they "apparently have been derived from a parent finch stock." Thus there may indeed have been genetic change in overall size, shape, color, feeding habits, size and shape of the beak, and all the other characteristics that differ among species of finches; yet similar degrees of differences between monkeys and apes cannot be the result of evolution, because Gish is sure that apes and monkeys were created as different "kinds."

Biologists have recognized ever since the dawn of taxonomy that the categories into which animals and plants are classified are arbitrary. Most of the higher categories in the classification of plants and animals are bridged by intermediate forms, so that the limits of each category are almost invariably ill defined. For example, the garter snakes and water snakes have been placed in separate genera, but a spectrum of semiaquatic species, differing slightly in almost every possible respect, connect them. Cobras are put in a separate family, but the fangs and poison glands that distinguish them are developed to varying degrees in certain other snakes. Snakes as a group are distinguished from lizards by their lack of legs and certain features of their teeth and jawbones, but among the lizards there are many species that have diminutive legs or none at all, and others with snakelike jaws. One group of species, the blind snakes, has been classified as lizards by certain taxonomists. Modern reptiles are all easily distinguished from amphibians by their skeletons, but Permian fossils such as *Seymouria* cannot be classified unambiguously as either reptilian or amphibian. So it goes, throughout the whole taxonomic hierarchy. Even the plant and animal kingdoms cannot be distinguished when you examine one-celled organisms that are claimed by the zoologists as protozoa and by the botanists as algae.

In Darwin's time, these gradations between groups were already recognized, but there was one smaller taxonomic category that was thought to be real—the species. However, Darwin pointed out that naturalists have had great difficulty in deciding whether similar forms should be classified as species or merely as varieties. When a fauna or flora is poorly known, he said, the species will seem distinct; but as it is studied more extensively, variations connecting the species are found, and the taxonomist becomes increasingly perplexed. Thus "no clear line of demarcation has as yet been drawn between species and sub-species —that is, the forms which in the opinion of some naturalists come very near to, but do not quite arrive at, the rank of species. . . . These differences blend into each other by an insensible series; and a series impresses the mind with the idea of an actual passage."[9]

One of the ironies of the history of biology is that Darwin did not really explain the origin of new species in *The Origin of Species*, because he didn't know how to define species. The *Origin* was in fact concerned mostly with how a single species might change in time, not how one species might proliferate into many. The modern idea of how speciation may happen, how one species can split into two or more species, was developed largely in the 1930s and 1940s, and chiefly by Ernst Mayr, who sought to clarify the definition of species.[10]

Mayr argued that two forms cannot be classified as different species merely by how different they look. In North America, for example, a small white goose known as Ross's goose is very similar, except in size and some slight differences in color, to the snow goose. The snow goose, however, is classified as the same species as the "blue goose," a dark gray bird with a white head. The blue goose and the snow goose are the same species because they interbreed freely: a pair of parents often have offspring of both types, as well as some that are intermediate in coloration. The blue and snow geese are members of the same gene pool, whereas the Ross's goose is a different gene pool. Ross's geese do not interbreed with snow geese, despite the great similarity between the species. Mayr therefore proposed the "biological species concept," which has won general acceptance: two forms, whether they differ greatly or only slightly, are different species if they have separate gene pools. The biological importance of this concept is that if species do not interbreed, each is free to develop different adaptations—to strike off on a separate evolutionary path. As long as two forms do interbreed freely, they cannot become more and more distinct, because any new genetic differences between them become mixed together.

The key difference between two related species, then, must be some characteristic that tends to prevent them from mixing their genes to-gether—a "reproductive isolating mechanism." There are several kinds of isolating mechanisms. One kind, called "postmating isolation," is exemplified by the mule. A horse and an ass can and often will mate with each other, but the hybrid, the mule, is sterile, so that it cannot serve as a bridge to transmit horse genes into an ass population or vice versa. When hybrids are sterile or, in some cases, incapable of surviving at all, it is because of incompatible differences in their chromosomes or genes. The two sets of genes may spell out different instructions for development, which conflict with each other. For example, the hybrid between northern and southern leopard frogs doesn't develop normally, apparently because the two sets of genes are adapted to different temperatures, and development gets out of kilter when one set instructs the embryo to develop quickly and the other instructs it to develop slowly.

Many species, however, are quite capable of having perfectly healthy, fertile hybrids, but simply don't do so in nature because they don't mate with each other in the first place. Among ducks, for example, females respond only to the particular color patterns and courtship displays of their own species. It is easy to obtain fertile hybrids between mallards and pintails in zoos, but they rarely occur in nature. Differences in the form of flowers, the mating calls of frogs, the sexual scents of moths, and the flash patterns of fireflies similarly act as barriers to interbreeding. These are called "premating isolating mechanisms."

Species split into new species, and give rise to greater diversity, when new isolating mechanisms arise. This seems to happen most often when different populations of a species, occupying different localities, gradually accumulate genetic differences that make the two forms incapable of interbreeding, or unwilling to. A good analogy to this process lies in the evolution of languages. We know, for instance, that the various romance languages are derived from Latin. Gradual changes in speech patterns occurred independently in Spain, France, and Italy, and by the ninth century the languages were different enough that although an Italian could probably have understood a Frenchman, it would have been difficult. By the nineteenth century, their speech had become mutually incomprehensible. There is no point at which one could have said they were different languages rather than mere dialects—they gradually became more and more different. If we were to define languages as mutually unintelligible tongues, then Italian and French are definitely different languages. In the same way, if we think of the genes as providing messages for how to develop, or how to mate, the genetic differences between populations are at first like those between dialects, posing a slight barrier to mutual genetic interchange. In time, they grow into "language" differences that cut off genetic interchange completely. If two isolated populations come into contact before this process has gone too far, they may interbreed where they meet, and remerge into one species. If the process has gone further, they may coexist without interbreeding and losing their separate identities.

If this hypothesis is true, different populations of a species should differ from one another to a greater or lesser degree, in the characteristics that affect reproduction, just as the dialectical differences between British populations, made famous by *Pygmalion* and *My Fair Lady*,

FIGURE 25. Until recently, the Baltimore oriole (top) in eastern North America, and the Bullock's oriole (bottom) in western North America, were isolated by the Great Plains. They now meet in the central United States, where hybrids such as the one illustrated by the central figure occur. The eastern and western forms had diverged almost to the status of separate species, but not quite fully. The populations differ in the pattern of black and orange, and in the white markings on the wing. (Redrawn from G. M. Sutton, *The Auk* 55:1 [1938].)

range all the way from slight regional accents to differences that prevent mutual understanding altogether. This is exactly the case. In eastern Asia, for example, populations of the gypsy moth from different parts of Japan and Korea differ in genes that affect sexual development. When moths from different populations (say, northernmost and southernmost Japan) are crossed, their offspring are sterile intersexes. Clearly, then, if these populations expanded and came into contact, they would remain separate gene pools—separate species—because the hybrids would be sterile. If, however, you cross moths from populations that are closer together, say from northern and central Japan, the offspring are not entirely sterile. The degree of sterility varies, depending on which populations are crossed.[11] The most plausible interpretation is that populations in different localities have developed greater or lesser genetic differences which influence the degree to which they can exchange genes when given the opportunity.

Populations from different localities also vary in their mating preferences. The geneticists Theodosius Dobzhansky and Lee Ehrman have studied mating in the laboratory between strains of the fruit fly *Drosophila paulistorum* that were collected in various parts of South America. Flies from the headwaters of the Amazon Basin in Colombia will not readily mate with flies from Guyana, but both of these strains will mate readily with flies from northern Venezuela. Thus, like the other characteristics of species, mating preferences evolve to be different in different geographic populations, and span a complete gamut from free interbreeding to complete isolation. It is obviously arbitrary in cases like the South American fly whether one classifies all the populations as one species or many. The populations are at least on the road toward becoming different species.[12]

In previous chapters, I described cases in which adaptive genetic changes such as insecticide resistance or increased body size have been observed both in the laboratory and in nature. Such changes don't result in new species, but only a transformation of the original species, unless the population's willingness or ability to interbreed with the rest of its species is altered. But alterations in the ability to interbreed have also been observed many times in the laboratory, showing that genetic changes leading toward speciation can happen rather quickly. For example, Theodosius Dobzhansky and Olga Pavlovsky reported[13] in 1971 that a strain of fruit flies they collected in Colombia was at first fully interfertile with a strain collected in the Orinoco Basin. They kept the two stocks separate for about five years, and then crossed them again. This time the male offspring were completely sterile. A substantial amount of genetic isolation had developed in only five years.

A different experiment has been reported by several workers who have divided a group of flies into separate populations and then selected the populations so that they developed differences in bristle number or heat tolerance. After about twenty generations, flies that have come to differ in these characteristics do not interbreed freely with one another when they are put together. For some reason, females come to prefer males of their own kind. It appears, then, that differences in mating preference may arise as a by-product of other genetic changes that transpire in populations while they are adapting to different environments.

No doubt evolution is usually quite a bit slower in the wild, but a great deal of evidence from nature also indicates that speciation can occur rapidly. In the Hawaiian Islands, for example, there is a group of moths found nowhere else. One of the species feeds on a Hawaiian species of palm. Five other species feed only on banana. But until about a thousand years ago there were no banana plants in Hawaii. They were brought there from Polynesia, where the moths do not occur. So these species must have evolved in the last thousand years.[14] Another example is that of some cichlid fishes that are found only in a small lake in Africa separated by a low sandbar from Lake Victoria, where related species of fish live. Carbon dating of charcoal fragments in the sandbar has indicated that the sandbar is probably only 4,000 years old, suggesting that the species in the small lake have evolved in just a few thousand years.[15]

In the cichlids and many other cases, new species seem to have been formed when a small population became isolated. Many biologists feel it is under such circumstances that the rapid evolution of new species is most likely to occur. The environment of a restricted locality—say, the pine barrens of central New Jersey—is often very different from the average environment over a much broader area, so that a small, localized population may experience very different environmental pressures that favor rapid genetic change. For example, many new species of fruit flies have developed on the island of Hawaii, which is less than 750,000 years old. Many of these species have very small populations, restricted to small patches of forest surrounded by desolate lava flows.[16]

In contrast, two very widespread populations that are separated by a barrier may not diverge very rapidly, because the environment may be quite similar, on average, over two large areas. The European and American sycamores, for example, have been separated for millions of years but are still almost identical and can hybridize easily. Thus it seems likely that populations undergo the most rapid genetic change, and form new species most rapidly, when they are small and inhabit a small area. The entire set of populations that make up a widespread species isn't likely to evolve very fast, because few environmental changes are likely to affect a whole continent; but a single population of that species, isolated in a valley somewhere, is quite likely to encounter peculiar local conditions and adapt to them rapidly. New species are therefore likely to appear alongside the species they evolved from, while the ancestor retains the species' original characteristics with little alteration.

This is the vision of evolutionary change that has been dubbed by Niles Eldredge of the American Museum of Natural History, and Stephen Jay Gould of Harvard University, "punctuated equilibrium." Eldredge and Gould point out[17] that if evolution works this way, the fossil record will be a history of sudden changes, interspersed by long periods in which not much happens. The ancestral, wide-ranging species doesn't change much; but in local areas it spawns new species that evolve new characteristics very rapidly. When they become reproductively isolated from their ancestor, they can quickly spread out and coexist with it—and so "suddenly" appear in the fossil record.

Not all evolutionists agree with Eldredge and Gould's emphasis on the role of speciation in evolution. Many feel that rapid evolution can occur not just when species split into two, but also when a single, undivided species adapts rapidly to shifts in the environment. Nevertheless, it is clear that when species proliferate, radically new forms can arise in a short time. Some of the fruit flies on the island of Hawaii differ greatly from their relatives on the other Hawaiian Islands. The islands also have a group of peculiar birds, the honeycreepers. Color patterns, chromosomes, and other lines of evidence show that all these species are closely related, yet they differ among each other more than some whole families of birds or flies differ from other whole families. Among the honeycreepers, for example, some have long slender curved bills for probing into flowers, while others have stout parrotlike bills adapted for crushing large seeds.[18] Many whole families of birds, such as warblers and finches, are distinguished by lesser differences in their beaks.

In several small bodies of water in the Death Valley area of California there exist related species of pupfish that differ in such characteristics as the presence or absence of pelvic fins—a characteristic that distinguishes some whole families of fishes. These species have developed since the last glacier receded ten to thirty thousand years ago, when the area became a desert.[19] It is clear, then, that when new species evolve, they can differ from each other to a great extent, or hardly at all. Some species of desert pupfishes are almost identical to each other, and share

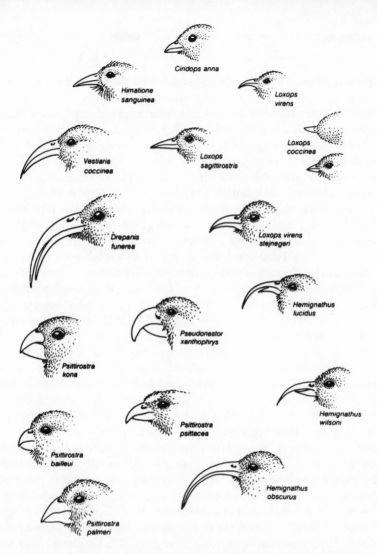

FIGURE 26. Adaptive radiation in a group of closely related species, the Hawaiian honeycreepers. As in the Galápagos finches (Figure 1, Chapter 1), the shape of the bill is an adaptation to the species' feeding habits. Species with short thin bills feed primarily on insects, those with long slender bills on nectar, and those with thick bills on fruit and seeds. Intermediate forms connect many species, such as *Psittirostra*, *Hemignathus*, and *Ciridops*, that are as different in bill shape from one another as are other families of birds (such as finches, hummingbirds, and tanagers). (From D. J. Futuyma, *Evolutionary Biology* [Sunderland, Mass.: Sinauer Associates, Inc., 1979]; after W. J. Bock, *Evolution* 13:194–211 [1970].)

many of the same genetic variations that are revealed by studying their proteins; others differ in size, tooth shape, coloration, and the presence or absence of fins. Among pupfish, there is a complete gradation from species that can hardly be distinguished to those that are as different as other genera or families. Moreover, every characteristic that can be used to distinguish families or orders is known to vary among closely related species, or even within species.

In his book *Flowering Plants: Evolution Above the Species Level,* G. Ledyard Stebbins tabulates some of the characteristics that distinguish families of plants yet vary among related plants elsewhere in the plant kingdom.[20] Whether a plant is woody or herbaceous, for example, is the key feature that distinguishes the Primulaceae (primroses) from the Myrsinaceae; the same distinction is the difference between the monkey-flowers *Mimulus longiflorus* and *Mimulus clevelandii*. Compound versus simple leaves separate the Oxalidaceae (wood sorrels) from the Linaceae (flax), and also the buttercup *Ranunculus repens* from *Ranunculus cymbalaria*. One of the most important features used to classify plants, whether the petals are fused to the ovary or not, separates the Rubiaceae (coffee family) from the Loganiaceae (butterfly bush), and also the saxifrages *Saxifraga umbrosa* and *Saxifraga caespitosa*. The point is that there is no mysterious "essence" that goes into making a primrose or a coffee plant; the primrose family and the coffee family are just groups of species that share certain combinations of characteristics, every one of which can differ just as extensively between related species of plants that are otherwise very similar. The same differences occur even within species. The key difference between two big subdivisions of the sun-flower family is the presence or absence of ray florets; but Jens Clausen and his collaborators[21] found a wild population of plants that, though they first thought it was a new species of one subdivision, turned out to be merely an aberrant population of a well-known species in the other subdivision. They crossed the aberrant and "normal" plants and found that the difference was due to just two genes. So in the hills of California a major evolutionary step may well be in the making.

If the characteristics that distinguish genera or families are the same as those that distinguish otherwise similar species; and if the differences among species can be found in the differences among local populations that cannot be clearly defined as the same or different species; and if the differences among populations consist of the same kinds of genetic variations that exist within populations; then there can be no such thing as discretely different "kinds." The transitions between populations and geographic races, between geographic races and species, between species and higher taxa are as gradual as are the transitions from regional accent to dialect to language to language group.

In fact, the evolution of life appears to be like the "evolution" of languages in many ways. Whether by chance, or because people adopt the speech of a peer group they wish to fit into (so that some forms of speech are perceived to be superior and are thus "selected"), slight changes in grammar and pronunciation develop from generation to generation. With time, if populations are separated from each other, dialectical differences become more pronounced and different languages develop. The differences in language are not due to genetic changes, of course, and for this reason peoples can quickly adopt each other's language when they come into contact. In this way linguistic evolution and biological evolution differ, but otherwise the analogy holds quite well. The analogy holds also for the methods by which linguistic evolution and biological evolution are studied. Linguists sometimes can use "fossils" to reconstruct the history of language—ancient inscriptions or poems, for example. But more often they have no written records on which to rely, and must use the "comparative anatomy" of languages —the differences and similarities of pronunciation and grammar—to reconstruct their history. In a way that is analogous to the biologist who turns to geological events like continental drift to confirm his suspicion about when certain groups of species diverged, the linguist turns to the history of wars and conquests to judge when major changes in language may have taken place.

Languages change gradually, and some become extinct; but no one doubts, despite the gaps, that the romance languages evolved from Latin, nor that Latin had a common ancestry with Sanskrit, the Germanic and Slavic languages, and the Celtic languages that included Old English.[22] No one has seen Latin give rise to Spanish, of course, nor are the historical documents adequate to trace the changes in detail. But direct observation is not the only source of knowledge, and it is often not even the best source; think of how often the "eyewitness accounts" in newspapers prove false. Knowledge proceeds, rather, from erecting hypotheses and testing whether or not the world conforms to the predictions that the hypotheses give rise to.

But this brings us to philosophical issues that are really the heart of the conflict between creationists and evolutionists. The issue at stake is not merely whether evolution has occurred or not. It is, rather, whether science is a reliable path to knowledge.

NINE

SCIENTIFIC KNOWLEDGE

I t is time for students of the evolutionary process, especially those who have been misquoted
and used by the creationists, to state clearly that
evolution is *fact*, not theory. . . . Birds arose
from nonbirds and humans from nonhumans.
No person who pretends to any understanding
of the natural world can deny these facts any
more than she or he can deny that the earth is
round, rotates on its axis, and revolves around
the sun.
—RICHARD C. LEWONTIN, 1981*

Before the mid-1930s, science in Russia could proudly hold its own
with any in the world. Within genetics and evolution alone, the names
Vavilov, Dubinin, Timofeef-Ressovsky, Severtzov, Schmalhausen, and
many others ranked with the best researchers in England, Germany, or
the United States. Then came Trofim D. Lysenko, a man unschooled

---

* Richard C. Lewontin, who holds the Alexander Agassiz professorship at Harvard
University, is a leader in research on the genetic basis of evolution.

in science but well attuned to political currents. Presenting not scientific evidence but Marxist rhetoric, he won the support of Stalin for his genetic theories. Within a decade the best scientists in the Soviet Union had been imprisoned, executed, or silenced.[1] Why?

According to Lysenko, Mendelism was "the tool of bourgeois society," particularly in holding that the gene could not be altered except by mutation, and that the direction of mutation was random—meaning that mutational change could not be governed by man or the environment. This, Lysenko believed, had to be wrong; because nature, as Marxist doctrine showed, had to be both improvable and perfectable. Adopting a kind of Lamarckism, Lysenko set out not only to destroy his Mendelian rivals but to transform Soviet agriculture. Improved varieties of crops could be created, he said, by allowing the environment to alter their hereditary properties. Within a single generation, he could transform winter wheat into spring wheat merely by changing the temperature in which it grew. He went further, saying that species could be transformed into other species at will—wheat into rye, for instance—by planting them in the right environment. The Darwinian idea that members of a species compete for the necessities of life was an invention of bourgeois science used to justify class struggle in a capitalist society. There could be no inherent competition, only altruism in nature. Seeds should be sown in clusters so that all except one would "sacrifice themselves for the good of the species."

Rapidly achieving command of Soviet biology and agriculture, Lysenko destroyed it all. From the 1930s on, biological research in the Soviet Union was perverted to Lysenko's ends, and agricultural production fell deeper and deeper into disaster. Lysenko had never given evidence for his claims; but his political power left him unassailable. Only in 1965 was he finally deposed, and Soviet biology has been struggling ever since to catch up with the West.

A grim story indeed, but what do we learn from it? That reality stubbornly refuses to be bent to our desires or ideologies. Genes cannot be altered to suit our ends, as devoutly as we may wish them to be. Truth cannot be established by the Communist Party, nor by the vote of a democratic society or a board of education. Reality does not yield to wishful thinking. There are certainly many people who believe in unicorns, or in the predictions of astrology, or in the transmigration of souls, because these beliefs satisfy their emotional needs. Such needs, however powerful, do not make unicorns or astrological influences or metempsychosis any more real. Nor do our wishes make unpleasant realities go away; death and disease remain grimly real.

In this light, how should we interpret such creationists' objections to teaching evolution as that it is harmful to the child because it "contradicts his innate consciousness of reality and thus tends to create mental and emotional conflicts within him"; it "tends to remove all moral and ethical restraints"; "it may tend to rob life of meaning and purpose"; it "leads to a conviction that might makes right"?[2] In other words, don't tell people about evolution because it's unpleasant, like death. Deny the reality of evolution and you can save the child from emotional conflicts and immorality.

Even if these accusations were true, they would be irrelevant to the question of whether or not evolution is a fact. Science is science only if it limits itself to determining the nature of reality. The hallmark of science is not the question "Do I wish to believe this?" but the question "What is the evidence?" It is this demand for evidence, this habit of cultivated skepticism, that is most characteristic of the scientific way of thought. It is not limited to science, but it isn't universal, either. Many people still cling to traditional beliefs in the face of contrary evidence, out of wishful thinking, the desire for security and simplicity. But rationalism, as the philosopher of science Karl Popper has said, "has . . . always claimed the right of reason and of empirical science to criticize, and to reject, any tradition, any authority, as being based on sheer unreason, or prejudice, or accident."[3]

At its best, science challenges not only nonscientific views but established scientific views as well. This, in fact, is the wellspring of progress in science. Our knowledge can progress only if we can find errors and learn from them. Thus, much of the history of science consists of a rejection or modification of views that were once widely held. Geologists once believed in the fixity of continents, but now believe in continental drift. The Newtonian theory of physics is now seen as a special case of a larger theory that includes relativity. Scientists realize, if they have any sense at all, that all their currently accepted beliefs are provisional. They are, at present, the best available explanations, but subsequent research may show them to be false or incomplete.[4] I cannot stress this point too strongly. Unlike fundamentalists who will not consider the possibility that they could be wrong, good scientists *never* say they have found absolute "truth." Read any scientific paper and you will find the conclusions couched in words like "apparently" or "it appears that."

Scientists accept uncertainty as a fact of life. Some people are uncomfortable unless they have positive, eternal answers; scientists come to terms with uncertainty and mutability as a fundamental condition of human knowledge. Science is not the acquisition of truth; it is the quest for truth.[5]

The picture I have just painted is, of course, a somewhat idealistic one. In fact, scientists are just as human as anyone else. They believe that one or another hypothesis is most likely to be true, and they engage in sometimes bitter battles to defend their ideas. Scientists' beliefs are also shaped by their political, social, and religious environment. It is undoubtedly true that Darwin and Wallace were led to the idea of natural selection because the English economic system of their day put an emphasis on competition, free enterprise, and economic progress. The history of IQ testing shows that scientists can often be misled by their social beliefs. Psychologists in the early part of this century "knew" that there were fixed hereditary differences between races in intelligence, and interpreted all the data they collected in this light.[6] One of the pioneers in IQ testing, H. H. Goddard, "discovered" by testing immigrants that 79 percent of the Italians, 83 percent of the Jews, and 87 percent of the Russians he tested were "feeble minded." The IQ test was in English, of course.

Thus the common image of scientists as abstracted, unbiased, detached intellects has no foundation in reality. Scientists are often highly opinionated, even in the face of contrary evidence; and they are often not particularly intelligent, either. The spectrum of scientists, as of any other group of people, runs from the brilliant to the fairly stupid. Almost every scientist has made more than one asinine statement in the course of his or her career, and some make them habitually.

If scientists can be just as biased, subjective, and foolish as anyone else, why should we have any belief in what they say about physics, evolution, or the causes of cancer? Because scientists are motivated not only by a quest for knowledge but a quest for reputation. And there is no better way for a scientist to achieve reputation than to demolish existing ideas by finding contrary evidence, or to propose a theory that explains the evidence better. This means that although individual scientists often make errors, the body of scientists in a field eventually uncovers these errors and attempts to correct them. Every scientist's research depends on the research of others in the field; so out of pure self-interest, every scientist scrutinizes the work of others carefully, to be sure it is reliable. Science is a self-correcting process.

How are errors found? First, most scientific journals send articles to other scientists for review before accepting them for publication. Such an article is expected to present not only experimental data but a detailed description of the methods by which the data were obtained, so that others can repeat the experiment. Reviewers scrutinize these papers and often reject them for any of several reasons: insufficient data, erroneous methods, improper use of statistics, unwarranted deductions from the data. About half the papers submitted to *Evolution*, the journal that I edit, are rejected for these kinds of failings.

Certainly not all errors are caught at this stage, but many of those that slip by are pointed out by other researchers in papers they subsequently publish. Any supposed fact or theory that is at all important is soon tested by other scientists, who see if they can confirm the results. Recently, for example, some investigators claimed that immune resistance acquired during an individual's lifetime could be passed on genetically. This claim of Lamarckian inheritance runs counter to genetics and evolutionary theory; if it were true, it would be extremely important. But other immunologists immediately tried to repeat the observations and were unable to verify them.[7] Such a claim will be held in abeyance, or rejected outright by the scientific community.

Very rarely, there are even cases of dishonesty. A recently publicized case is that of the late Cyril Burt, whose data on the IQs of separated pairs of twins provided the main evidence for a genetic basis for IQ variation. Burt's critics have now found inconsistencies in his data, and even his supporters have ultimately agreed that he must have made up data.[8] But such cases are extraordinarily rare, because every scientist worthy of the name knows that his or her data must withstand scrutiny.

The result of this process of exploration and correction is that at any time, scientists have a body of knowledge and understanding that is reliable, within the limits of what can be known at that time. It is for this reason that I, as a biologist, can have confidence in, say, radiometric dating, or in the atomic theory of chemical reactions, even though I have no personal knowledge of these fields. Radiometric dating is so critical to physics and geology that it would not be widely used unless its validity had been repeatedly tested. A scientist therefore does not appeal to the authority of any one scientist to justify his or her beliefs, but rather to the entire body of scientific practice.

So far, I have been talking about science without defining it. Many people assume that science is the collection and cataloging of facts, but science is much more than this. Darwin saw this clearly: "About thirty years ago there was much talk that geologists ought only to observe and not to theorize; and I well remember someone saying that at this rate a man might as well go into a gravel-pit and count the pebbles and describe the colours. How odd it is, that anyone should not see that all observation must be for or against some view, if it is to be of any service!"[9] That is, science consists of a search for explanations. The *Oxford English Dictionary* defines science as "a branch of study which is concerned with a body of demonstrated truths, or with observed facts, systematically classified and more or less colligated by being brought under general laws, and which includes trustworthy methods for the discovery of new truths within its own domain."

I have already described some of these "trustworthy methods." What makes them trustworthy is repeatability. An observation is accepted as a scientific "fact" only if it can be repeated by other individuals who follow the same methods. Thus extrasensory perception is not considered factual by most scientists, because so far it has not been possible for skeptical observers to verify the claims of people who say they have ESP.

But what are the "truths" that science is supposed to discover? "Truth," according to the same dictionary, is "conformity with fact; agreement with reality." "Fact," in turn, is "something that has really occurred or is actually the case; something certainly known to be of this character; hence a particular truth known by actual observation or authentic testimony." But whose knowledge or observation serves to establish something as fact? Certainly not the populace at large. It is very possible that most people in the world do not know the fact that the earth revolves around the sun. Then what kind of observation establishes a fact? Not direct observation, necessarily. We observe the sun moving across the sky, but not the fact of the earth's rotation. We accept the fact that material is made of atoms, but we have no personal experience of them. In the scientific sense, then, "facts" must be propositions agreed upon by individuals who have repeatedly applied rigorous, controlled methods of direct or indirect observation. All but the most trivial facts ("There is a blue chair in my office") begin life as hypotheses, and graduate to "facthood" as knowledgeable individuals come to agree upon them. The rotation of the earth was once a hypothesis; it is now a fact. Facts are merely hypotheses that are well supported by the available evidence.

The word "hypothesis" means, to many people, an ungrounded speculation. But this is not the way the word is used in science. As Sir Peter Medawar says,

> Most words of the philosopher's vocabulary, including "philosopher" itself, have changed their usages over the past few hundred years. "Hypothesis" is no exception. In a modern professional

vocabulary a hypothesis is an imaginative preconception of *what might be true* in the form of a declaration with verifiable deductive consequences. It no longer tows "gratuitous," "mere," or "wild" behind it, and the pejorative usage ("evolution is a mere hypothesis," "it is only a hypothesis that smoking causes lung cancer") is one of the outward signs of little learning.[9]

The difference between a fact and a hypothesis, then, is a matter of degree, a matter of how much evidence there is. Yet people who have a vested interest in opposing scientific conclusions often claim that whatever they oppose is merely a hypothesis or a theory, not a set of facts. For several decades there has been overwhelming evidence that smoking causes lung cancer; but the tobacco industry says that this hasn't been "proven," that the relationship is a "hypothesis" instead of a "fact." In exactly the same way, creationists say that evolution is a theory, not a fact, and hence unproven. However, nothing in science is ever proven in this sense. There are no immutable facts. Every scientific claim is a hypothesis, however well supported it may be. It has never been proven that hemoglobin carries oxygen; there is merely so much evidence for this claim that it is hardly imaginable that it could be false. Nonetheless, it is still conceivable that some revolution in chemistry could completely change the theory of chemical bonds, and that we would have to revise our notion of what hemoglobin does.

It is important to recognize that not all "facts" are susceptible to scientific investigation, simply because some observations and experiences are entirely personal. I cannot prove that someone loves his or her child. The emotions that any individual claims to have are not susceptible to scientific documentation, because they cannot be independently verified by other observers. In other words, science seeks to explain only objective knowledge, knowledge that can be acquired independently by different investigators if they follow a prescribed course of observation or experiment. Many human experiences and concerns are not objective, and so do not fall within the realm of science. As a result, science has nothing to say about aesthetics or morality. It cannot provide an objective basis on which to judge whether or not Beethoven wrote great music, or whether or not an act is ethical. The functioning of human society, then, clearly requires principles that stem from some source other than science. While science can provide objective knowledge, we must look elsewhere for guidance on how to use that knowledge.

Scientific facts are, as Darwin suggested, usually not very interesting unless they bear on theories that explain them. "Theory" in science has a very special meaning. It does not mean mere speculation or conjecture. Rather, as the *Oxford English Dictionary* puts it, a theory is "a scheme or system of ideas and statements held as an explanation or account of a group of facts or phenomena; a hypothesis that has been confirmed or established by observation or experiment, and is propounded or accepted as accounting for the known facts; a statement of what are known to be the general laws, principles, or causes of something known or observed." Thus relativity theory and Newtonian theory are bodies of interconnected statements that in combination explain physical events. Atomic theory is a body of statements about the structure of atoms that explains chemical reactions. Plate tectonic theory is a body of statements about forces operating within the earth that, among other things, are responsible for the movement of continents. Each of these theories relates a vast number of formerly disconnected phenomena. Plate tectonic theory, for example, brings together observations in seismology, geomagnetism, geochemistry, and various branches of terrestrial and marine geology. When, in this way, a theory makes sense out of many otherwise mysterious phenomena, it is likely to be accepted even before there is actually adequate evidence. Nevertheless, every such theory makes predictions, which then test the theory's validity. The Copernican theory of the solar system, although thoroughly accepted by the seventeenth century, was not definitively tested until the 1830s, when the predictions it made about stellar parallax (the direction of a star from different places on the earth's orbit) were verified.

One criterion of a good scientific theory is that it make predictions which, if subsequently confirmed, serve to support the theory. It must be kept in mind, however, that even if they support it, they do not prove it true. It is always possible that another theory, yet to be conceived, would make the same predictions. Thus there is another important criterion of a scientific theory which most scientists accept. This is Karl Popper's dictum that a theory be in principle "falsifiable."[10] That is, a good theory doesn't merely explain everything; it specifically predicts that certain observations, if made, would prove the theory wrong. If you propose that diseases are caused by evil spirits, there is no way in which I could possibly prove your theory wrong. Such "spirits" would not be detectable; and if they act at whim, I can make no predictions about who will or will not fall sick. The germ theory of disease, on the other hand, makes specific predictions about infectivity, physical conditions that favor the contraction of disease, and so on. It predicts that a disease could not be caused by germs if people suffer it who have never contacted infected material.

The most powerful form of science, then, consists of formulating hypotheses, sometimes by observation and sometimes by intuition, analogy, or other sources of insight that we do not fully understand; and deducing conclusions from these hypotheses that can be tested directly or indirectly by observation or experiment. The testing of hypotheses is the "body of trustworthy methods" to which the *Oxford English Dictionary* refers.

If a scientific theory is one that can be corroborated by observations that accord with its predictions, that can be falsified by observations or experiments which are incompatible with the theory, and that relies on objective observations that can be repeated by trained, unbiased observers, a nonscientific theory must be the converse. The nonscientific theorist lives within an impregnable fortress, safe from criticism, because the hallmark of nonscientific theories is that they cannot be falsified. They are formulated vaguely, or invoke agents whose actions cannot be predicted, so that they "explain" every possible outcome of a situation. Whatever your personality or history may be, a good astrologer will find some conjunction of the planets that explains why you are this way, even though as a Sagittarius you're "expected" to be the opposite.

Similarly, any "theory" that explains phenomena by recourse to the actions of an omnipotent, omniscient supreme being, or any other supernatural omnipotent entity, is a nonscientific theory. I could postulate that all human actions are slavish responses to the suggestions of guardian angels and diabolical incubi, and no one could possibly prove me wrong; for whether a person's actions look rational or irrational, good or evil, I can involve the power of supernatural suggestion. I could similarly postulate that God personally has governed the development and life of every creature that has ever been born, and if you protest that the laws of physics, chemistry, and biology explain biological phenomena, I could answer that God in his wisdom sees fit to act in an orderly way that gives the appearance of material laws of causation.

Because such a theory cannot be challenged by any observation, it is not scientific. It isn't necessarily wrong. It is just not amenable to scientific investigation. Science cannot deny the existence of supernatural beings. It cannot prove that God didn't create the universe. It can't prove that God isn't shunting electrons down the cytochrome molecules in your mitochondria. Science can neither affirm nor deny supernatural powers. Science is the exercise of reason, and so is limited to questions that can be approached by the use of reason, questions that can be answered by the discovery of objective knowledge and the elucidation of natural laws of causation. In dealing with questions about the natural world, scientists must act as if they can be answered without recourse to supernatural powers. There can be no scientific study of God.

How do these generalizations about science apply to the problem of evolution? We are dealing with two distinct questions. The first is the historical question of whether or not evolution has actually occurred: Have living forms actually descended by common ancestry from earlier forms? The second question is: If evolution has actually happened, what mechanisms have been responsible for it?

Both these questions have traditionally been subsumed under the term "the theory of evolution." But I wish to distinguish them carefully, for I consider the first question to have been resolved into fact, and the second question to fall into the category of theory.[11] The body of statements about mutation, natural selection, genetic drift, and so forth is the theory of evolution: that is, the explanation of the historical fact that evolution has actually occurred. The mass of evidence from the geological record, embryology, comparative morphology, biochemistry, and the rest of biology indirectly proves the common ancestry of living organisms to the satisfaction of biologists generally. This may seem an elitist basis on which to judge a proposition factual, but it is no more so than the elitism we accept in astronomers or, for that matter, in physicians who attribute disease to germs rather than to spirits.

Why do biologists consider evolution to be a fact? Partly because the hypothesis of evolution is corroborated by enormous masses of consistent evidence, just as untold numbers of astronomical observations corroborate the Copernican rather than the Ptolemaic view of the solar system. Every anatomical or biochemical resemblance between species, every vestigial structure, every pattern of geographical distribution, every fossil is consistent with the idea of evolution. Again and again, new discoveries, such as the close resemblance of the DNA of humans and apes, accord with the idea of common ancestry. But there is more to a good scientific hypothesis than corroboration; it must be falsifiable. And the hypothesis that evolution has occurred could indeed be falsified. A single undisputed fossil of a flowering plant or of a human or any other mammal in Precambrian rocks would do it. Millions of conceivable paleontological discoveries could disprove evolution, but none ever has come to light.

Unlike the historical fact of evolution, which is universally accepted by qualified biologists, evolutionary theory, that is, the theory of mutation, recombination, natural selection, genetic drift, and isolation, is subject to argument, just as there is argument about how genes are regulated during development, or how earthquakes are caused. Two major kinds of argument about evolutionary theory occur within scientific circles. There are philosophical arguments about whether or not evolutionary theory qualifies as scientific theory, and substantive arguments about the details of the theory and their adequacy to explain observed phenomena.

It is possible to ask whether or not a theory based on historical events can be tested, since such events are not susceptible to experimental manipulation or direct observation. However, if we cannot accept the idea that theories of history can be tested, most of the problems studied by scientists immediately cease to be amenable to scientific inquiry; because with the exception of some principles of physics and chemistry, most phenomena must be explained in part by recourse to history. Much of geology and astronomy, for example, deals with historical phenomena. Virtually all of biology is historical. If we ask why the forests of Long Island are dominated by pine instead of maple, the immediate answer is that the dry sandy soil of Long Island is more favorable to pines than maples; but to be fully satisfied, we must go on to ask why Long Island has sandy soils. The answer is, of course, historical: Long Island is a pile of loose rubble deposited by the most recent glacier, and does not have a mineral soil formed from underlying bedrock.

In fact, historical phenomena can be scientifically analyzed because they form patterns, showing that repeated historical events are generally associated with one or more conditions that imply causation. It is difficult, and often impossible, to rigorously test a hypothesis about any single historical event, such as why the human species is the only "naked ape," and most speculations about such unitary historical events must remain speculations rather than rigorous scientific statements. But when a particular kind of historical event is repeated, we see whether it is usually associated with specific conditions that might qualify as causes.[12] For example, from fossil material and comaprative anatomy, it is possible to test the hypothesis that new features of organisms generally evolve by modification of preexisting features.

A secondary issue then arises: Is the hypothesis of natural selection falsifiable, or is it a tautology? If there were no more to the theory of evolution than that "natural selection is the survival of the fittest," and we then defined the fittest as those best capable of survival, natural selection would indeed be an empty, untestable concept. The claim that natural selection is a tautology is periodically made in the scientific literature itself, and in fact had been made by Karl Popper, the chief advocate of testability in scientific theory. However, Popper has recently stated that he does now believe natural selection to be testable: "The theory of natural selection may be so formulated that it is far from tautological. In this case it is not only testable, but it turns out to be not strictly universally true."[13] In fact, the notion of natural selection has been tested many times. Thus, knowing that birds generalize color patterns from one insect to another, one can predict and then experimentally demonstrate that if an edible insect resembles an inedible species it will enjoy some protection from predation, and that the degree of protection increases with the degree of similarity to the unpalatable species.[14]

The neo-Darwinian theory of evolution is also clearly falsifiable because we can postulate alternative theories which, if true, would render neo-Darwinian theory superfluous. The most obvious alternative theory is the Lamarckian one. If it were true that modifications acquired during the life of an organism could become hereditary, many features of organisms would evolve by the direct influence of the environment, and natural selection would not play a major guiding role in adaptation.

These are the larger philosophical questions that can be asked about the validity of evolutionary theory, and most biologists are satisfied with the answers. Biologists do not universally agree, however, that the neo-Darwinian theory is *sufficient* to explain all of evolutionary change, and there is a lot of debate about which of the mechanisms of evolution are most important. For example, some of the most eminent evolutionists argue that not all evolution can be attributed to natural selection. Much of evolution may proceed by genetic drift, so that not all the differences among species are necessarily unique adaptive solutions to unique adaptive problems. Moreover, they argue, mutations are not random, in that certain kinds of mutations are more likely than others. However, almost everyone agrees that the chance of occurrence of a mutation is not influenced by whether or not the current environment would favor it.

Another major area of debate is whether or not mutations of small effect are the sole stuff of evolution. Although much of evolution certainly has proceeded by gradual changes, it is possible to imagine beneficial mutations that could produce large changes. Thus some morphologists and paleontologists feel that some major changes in evolution may have entailed "macromutations" with large effects. But neo-Darwinian theory does not invoke any natural law that mutations must be small in effect. The reason for supposing that most evolutionary change has been gradual is not theoretical but empirical—the observation that most variation within and among closely related populations is due to many genes, each with a small effect.

Finally, paleontologists such as Stephen Jay Gould have written that neo-Darwinian theory is insufficient to explain the broad panorama of historical evolution.[15] By "neo-Darwinian theory" he means the "hard-core" genetic theory that many evolutionists adhered to after the "modern synthesis" of the 1940s and 1950s: the belief that all of evolution consisted of the action of natural selection on slight genetic variations. If this is the definition of neo-Darwinian theory, then as Gould says, "the modern synthesis is incomplete, not incorrect." For such a theory would not account for nonadaptive characteristics; nor, in itself, would it account for why certain long-term series of evolutionary events occurred. In particular, Gould and certain other paleontologists have postulated that long-term evolutionary trends may be caused not just by slow, steady change in one direction within a particular species, but by a higher-level process that the neo-Darwinists didn't emphasize in the modern synthesis: rates of extinction and speciation.

For example, if the lineage from the eohippus to the modern horse increased in overall size for 60 million years, it is possible to imagine that this was the result of the steady, excruciatingly slow change of a single species. But the evolution of horses, as described in Chapter 4, included a lot of speciation: species continually proliferated new species, some larger and some smaller. Paleontologists such as Niles Eldredge, Stephen Jay Gould, and Steven Stanley favor this interpretation: new species, with various body sizes, arise quite quickly. The changes in body size occur by the processes of genetic variation and natural selection. But larger species may tend to survive longer than small species before becoming extinct. As a result, they have more of a chance to give rise to more large species than small species will to small species. Therefore, various species at any time may be evolving in both directions, but an overall trend toward larger size emerges, just because of differences in extinction rates. Such rates are not normally taken into account in the neo-Darwinian theory of natural selection within species. If the Eldredge-Gould-Stanley viewpoint is correct, we need a theory of why some species are more prone to extinction than others, to complement the genetic theory of natural selection.

The reason I am dwelling on this subject is that creationists have gleefully pounced upon the writings of these paleontologists, claiming that the whole structure of evolutionary theory is going down the drain. According to Gish, "Evolutionists are saying . . . that natural selection has made no significant contribution at all to the overall course of evolution."[16] But this is not at all what the paleontologists are saying. They are merely arguing that "macroevolution," the long-term history of life, includes important events, such as extinction, that cannot be studied in the context of "microevolution," the genetic changes of single species. Thus Gould argues that "macroevolution has some claim to theoretical independence": that understanding the history of life requires more information than genetic studies can provide, a whole level of theory that incorporates the genetic theory and adds to it.

Debates and controversies of this kind go on continually in evolutionary science and in every scientific discipline. They indicate not that the field is tottering on the brink of chaos and despair, but that scientific inquiry is flourishing: that people have found unexplored questions to answer, and new, improved theories with which to expand the scope of human understanding. New ideas come up continually and either pass the gauntlet of scientific testing or pass into oblivion. How different science is from creationism! Creationists, by their own admission, cannot test their theory. According to Gish, animals and plants "were brought into existence by acts of a supernatural Creator using special processes which are not operative today."[17] How, then, could they be examined by any methods of science? Gish goes on: "We do not know how the Creator created, what processes He used, *for He used processes which are not now operating anywhere in the natural universe.* . . . We cannot discover by scientific investigations anything about the creative processes used by the Creator" (emphasis in the original).[18] Exactly because it is impossible scientifically to investigate supernatural processes, creationists can offer no more evidence for creation now than they could in 1859. They simply repeat the same arguments they have used for centuries—no new ideas, no new information. Where science frees and exercises the human intellect, creationism claims the intellect is powerless. Where science offers the optimism that comes with understanding, creationism denies it. Where science grows, creationism stays stagnant. Where science offers the method of hypothesis-testing to justify its claims, creationism offers blind faith in the authority of a single book and its most rigid interpreters.

TEN

CREATIONIST
ARGUMENTS

owever much the creationist leaders might hammer away at their "scientific" and "philosophical" points, they would be helpless and a laughing-stock if that were all they had. It is religion that recruits their squadrons. Tens of millions of Americans, who neither know nor understand the actual arguments for—or even against—evolution, march in the army of the night with their Bibles held high. And they are a strong and frightening force, impervious to, and immunized against, the feeble lance of mere reason.
—ISAAC ASIMOV, 1981

A few years ago, the physicist Luis Alvarez and his colleagues proposed an extraordinary hypothesis to explain the extinction of the dinosaurs and many other forms of life at the end of the Cretaceous period. They suggested the cause was an asteroid's collision with the earth. The asteroid's impact, it was imagined, might have spread so dense a pall of dust through the atmosphere that the sun's rays could not penetrate and most plants (hence many animals) were extinguished.[1] At first glance, this idea seems to violate uniformitarianism, but the Cretaceous extinction was an extraordinary event, and collisions of this kind are, in fact, perfectly natural. Large meteorites strike the earth fairly often, and the craters of the moon show that massive impacts have altered that body's landscape. Astronomical calculations indicate that a major collision could well have happened quite a few times in the 4 billion years of the earth's history.

A hypothesis, however, requires more than plausibility to be adopted; it requires evidence. In upper Cretaceous strata, Alvarez and his colleagues found such evidence in the form of a widespread, thin deposit of iridium and certain other elements that are rare on earth but more abundant in meteorites and other bodies in the solar system. This was still not enough to establish the hypothesis. Like any good scientific theory, this one contains the germs of its own destruction. If the collision hypothesis is true, most of the extinctions at the end of the Cretaceous should have occurred within a few hundred years, at most. The hypothesis thus suggests a program of research: find out how concentrated the extinctions were in time. The answer is not yet in, but paleontologists looking into the matter seem to find that the extinctions were spread over a much longer period than the collision hypothesis requires, so it is beginning to look as if the theory may not prove viable.[2]

In contrast, when we turn to creationist literature, we find no such process of investigation and testing at work. In actuality, almost all creationist literature simply consists of attacks on evolution, rather than positive evidence for creation. To the creationists, any evidence against evolutionary theory apparently constitutes evidence in favor of creation. Their consistent tactic is to present as "scientific evidence for creation" what are no more than criticisms of existing scientific theories. An especially instructive example is the pamphlet "Summary of Scientific Evidence for Creation," by Duane Gish and his colleagues.[3]

From an evolutionist's point of view, the pamphlet makes two major claims. The first is that "life was suddenly created." The "evidence" for this statement reads in its entirety as follows:

> Life appears abruptly and in complex forms in the fossil record, and gaps appear systematically in the fossil record between various living kinds. These facts indicate that basic kinds of plants and animals were created.
>
> The Second Law of Thermodynamics states that things tend to go from order to disorder (entropy tends to increase) unless added energy is directed by a conversion mechanism (such as

photosynthesis), whether a system is open or closed. Thus simple molecules and complex protein, DNA, and RNA molecules seemingly could not have evolved spontaneously and naturalistically into a living cell; such cells apparently were created.

The laboratory experiments related to theories on the origin of life have not even remotely approached the synthesis of life from nonlife, and the extremely limited results have depended on laboratory conditions that are artificially imposed and extremely improbable. The extreme improbability of these conditions and the relatively insignificant results apparently show that life did not emerge by the process that evolutionists postulate.

In fact, complex forms of life (for instance, invertebrate animals) are preceded by a long fossil history of simple one-celled organisms; both the organic and inorganic world is rife with chemical mechanisms like crystals that produce order from disorder; and laboratory experiments have repeatedly demonstrated the natural assembly of the macromolecules that are the constituents of life, under conditions that are not at all improbable. But even if we put aside this evidence for the moment, the point is that this quotation consists entirely of negative statements, not positive evidence for the sudden creation of life.

The second anti-evolutionary claim for which this pamphlet gives "evidence" is that "all present living kinds of animals and plants have remained fixed since creation, other than extinctions, and genetic variation in originally created kinds has only occurred within narrow limits." Here, again in its entirety, is the supporting statement:

Systematic gaps occur between kinds in the fossil record. None of the intermediate fossils that would be expected on the basis of the evolution model have been found between single-celled organisms and invertebrates, between invertebrates and vertebrates, between fish and amphibians, between amphibians and reptiles, between reptiles and birds or mammals, or between "lower" mammals and primates. While evolutionists might assume that these intermediate forms existed at one time, none of the hundreds of millions of fossils found so far provide the missing links. The few suggested links such as *Archaeopteryx* and the horse series have been rendered questionable by more detailed data. [No reference is given.] Fossils and living organisms are readily subjected to the same criteria of classification. Thus present kinds of animals and plants apparently were created, as shown by the systematic fossil gaps and by the similarity of fossil forms to living forms.

A kind may be defined as a generally interfertile group of organisms that possesses variant genes for a common set of traits but that does not interbreed with other groups of organisms under normal circumstances. Any evolutionary change between kinds (necessary for the emergence of complex from simple organisms) would require addition of entirely new traits to the common set and enormous expansion of the gene pool over time, and could not occur from mere ecologically adaptive variations of a given trait set (which the creation model recognizes).

The second paragraph of this quotation is, of course, a mere assertion, given without any evidence. The first paragraph is full of misinformation. Many fossil groups *are* very different from any living organisms, and *are* classified into their own taxonomic categories; not only the familiar dinosaurs, but graptolites, trilobites, placoderms, multituberculates, condylarths, and many other less familiar groups. Recent work has not rendered questionable the evolutionary status of *Archaeopteryx* or the horse series. There are many fossils that show various degrees of intermediacy between the various groups that the authors claim are not connected by intermediates. But more important, where in this "Summary of Scientific Evidence for Creation" is the scientific evidence for creation? Neither this tract nor any other creationist literature presents new information, or ideas for research that could resolve the issue in favor of creationism.

To analyze creationist literature is to scale a fortress of facts and quotations taken from the evolutionary literature, distorted and quoted out of context, haphazardly glued into a defense around their faith. It is necessary to extract from this literature the major themes, the rocks that make up the edifice, and to see what kind of cement is used to keep the structure from crumbling.

The cement is, for the most part, rhetoric, the tool of the Sophists, who taught their pupils how to win arguments, rather than how to seek for truth. Creationist rhetoric includes many tactics. One is George Orwell's "Newspeak"—hiding the true nature of one's position under a name that implies its opposite. Just as "war is peace" in *1984*, creationism is "science" in the creationist literature (despite their own admission that it cannot be science).[4] The creationist relies on "the scientific law of *cause and effect*,"[5] we are told. This "law" adds a patina of scientific respectability to their efforts without being used to derive any scientific conclusions. Since the words "hypothesis" and "theory" are used in science to imply testability, the creationists don't use them. Instead, they apply to both evolution and creation the scientific-sounding word "model" in order to conceal the fact that evolution is a testable hypothesis, whereas creation is not.

Another creationist tactic is the profuse citing of documentation, without distinguishing between qualified and unqualified sources. Thus Gish cites prominent evolutionists such as Theodosius Dobzhansky; biologists such as Pierre Grassé who carry on the French tradition of rejecting natural selection in favor of Lamarckism (for historical reasons of French philosophy and chauvinism);[6] and creationists such as Henry Morris, as if all were equally expert authorities on evolution. In support of the thesis that all living things, including humans, existed contemporaneously, *Scientific Creationism* quotes a report from *Science Digest* (which is not a scientific research journal):

> An ancient Mayan relief sculpture of a peculiar bird with reptilian characteristics has been discovered in Totonacapan, in the northeastern section of Veracruz, Mexico. José Diaz-Bolio, a Mexican archaeologist-journalist responsible for the discovery, says there is evidence that the serpent-bird sculpture, located in the ruins of Tajin, is not merely the product of Mayan flights of fancy, but a realistic representation of an animal that lived during the period of the ancient Mayan—1000 to 5000 years ago.

The evidence referred to is not discussed, but *Science Digest* suggests that the sculpture "bears a vague resemblance" to *Archaeopteryx*, which existed during the age of dinosaurs. On this extraordinarily flimsy basis, the writers of *Scientific Creationism* conclude: "The evidence seems clear that archaeopteryx, or some equivalent ancient bird, was contemporaneous with man and only became extinct a few thousand years ago."[7]

In addition to such flimsy citations, creationist literature regularly quotes prominent evolutionists totally out of context or in carefully edited snippets so that they seem to be supporting creationist positions. For instance, according to Henry Morris, Harvard geneticist Richard Lewontin, one of the world's outstanding evolutionary biologists, "has rejected the Darwinist concept of struggle and survival, even at the genetic level."[8] Lewontin's article in *Scientific American*,[9] to which Morris refers, shows no hint of such a claim, and doesn't address the "Darwinist concept of struggle and survival" at all. It is instead a cautionary essay on the problems of distinguishing adaptive from nonadaptive characteristics, and a description of how natural selection can alter the characteristics of a species without making it better adapted to the environment.

As we have seen, the creationists are particularly fond of pointing to controversies in evolutionary literature as evidence that evolutionists are revising their ideas in order to save themselves embarrassment. Thus in their view, the "punctuated equilibrium" hypothesis of Eldredge and Gould has been devised to explain gaps in the fossil record because it is "the only remaining alternative to creationism,"[10] though this couldn't be further from the truth.

The line between misusing interpretations and actually distorting facts is a fine one. Creationists claim, for example, that the chronological ordering of geological strata depends on an evolutionary interpretation of fossils from simple to complex, and that by circular reasoning the fossil record is then taken as evidence of progressive evolution.[11] This, however, is simply not true. As paleontologist David Raup says, "The geological time scale in its modern form was fully developed by about 1840—before Darwin's *Origin of Species*. The time scale based on fossils was built by geologists who were creationists. Since 1840, many details have been filled in, but the basic sequence has remained unchanged.[12]

At another level, for several years Duane Gish has used the bombardier beetle to illustrate his claim that complex adaptations cannot develop gradually by natural selection, and so must have been created. By a chemical reaction of hydroquinone and hydrogen peroxide, the beetle produces a puff of a noxious chemical that repels enemies. Gish has claimed that this is an explosive reaction that would blow the beetle up. According to him, the explosion is controlled by a complex system of inhibitors that cannot be explained by natural selection. Thomas Jukes, a biochemist at the University of California, tells the following interesting anecdote:

> W. Thwaites and F. Awbrey at San Diego State University invited Gish to debate this matter publicly, in 1978. They mixed hydroquinone with hydrogen peroxide. There was no explosion; the mixture turned brown. Gish then claimed that he had mistranslated the original German reference, but, in spite of this, he continued, in 1980, to recite the false story about this mixture spontaneously exploding. He also alleges that the dragons of mythology were probably dinosaurs that used the bombardier beetle mechanism to breathe fire.[13]

Two more weapons in the creationist arsenal of rhetoric bear mention. One is the standard rhetorical trick of caricaturing the opponent's position to make it look ridiculous; the other is the demagogic principle of appeal to emotion rather than reason. Any position can, of course, be ridiculed, and creationists are masters of the art. For example, I heard Gish give a lecture[14] in which he quoted from a *Scientific American* article to the effect that whales and porpoises are believed, on biochemical evidence, to have evolved from artiodactyl ancestors— the group to which cows, pigs, and antelopes belong. Of course the article implied that cows and whales have a common ancestor, which was neither cowlike nor whalelike. But Gish proudly showed a home-drawn cartoon of a Holstein cow transforming itself, by a series of clearly ridiculous stages, into a whale—and triumphantly announced that this slide makes evolutionists furious. I was indeed angry—not at the triumph of creationist analysis, but at such a blatant caricature of evolutionary principles. He could as well have read from a textbook that animals are descended from the same ancestors that gave rise to plants, and then shown a cartoon of an oak tree being transformed into a man.

The creationist appeal to emotion takes many forms, but none is more unjustified than their repeated attempts[15] to blame evolutionary science for racism, Nazism, and the ethics of self-interest. Social Darwinism, which almost all biologists now reject, is indeed a pernicious doctrine (see Chapter 12), but man's inhumanity to man didn't begin in 1859. Henry Morris tells us that the "modern harvest" of evolution is Nazism: "The philosopher Friedrich Nietzsche, a contemporary of Charles Darwin and an ardent evolutionist, popularized in Germany his concept of the superman, and then the master race. The ultimate outcome was Hitler, who elevated this philosophy to the status of a national policy."[16] Can anyone believe that racism and anti-Semitism needed an evolutionary rationale; that without Darwinism they would not have been served by other philosophies, as they had been for centuries? Can Morris be unaware of the generations of evolutionists, from Darwin to Dobzhansky, who have pleaded for human rights and celebrated human diversity and the brotherhood of man?

So much for the rhetorical glue on which the integrity of the creationist structure depends. What are the creationist arguments against evolution?

1. Fundamentalist creationism requires that the earth be only a few thousand years old. However, the creationists are faced with a vast body of opposing evidence, including the great thickness of many sedimentary deposits. Some single deposits are as much as twelve miles thick. Even allowing for the most rapid rates of deposition that might be caused by local floods, such a depth of sediments would take at least 32,000 years to accumulate—and these deposits are but a tiny fraction of the entire history recorded in the rocks.[17] Even at their fastest known rates, processes such as seafloor spreading, mountain building, and erosion must have taken many millions of years to form the features of the earth's crust. In fact, the evidence from radioactive dating, which provides absolute ages, shows that these other sources of evidence underestimate the age of the earth.

The scientific inferences from these data rest on the principle of uniformitarianism, the belief that past processes were the same as present ones, so that the present is the key to the past. Uniformitarianism does not mean absolutely constant rates: the delta at the mouth of a river, for example, builds up faster at some times than others. The creationist "model," however, "is fundamentally catastrophic because it says that present laws and processes are *not* sufficient to explain the phenomena found in the present world."[18] As "evidence" for catastrophism, the authors of *Scientific Creationism* cite cases in which fossils seem to have resulted from rapid death and burial. But the catastrophe suffered by an oyster bed when a river suddenly dumps sediment on it after a spring thaw is hardly a violation of uniformitarianism, and it doesn't begin to account for sedimentary rocks twelve miles thick.

Creationists claim that there is no objective way by which to determine the age of rocks. In *Scientific Creationism*, we read the following extraordinary statements: "*Rocks are not dated radiometrically.* Many people believe the age of rocks is determined by study of their radioactive minerals—uranium, thorium, potassium, rubidium, etc.—but this is not so. The obvious proof that this is not the way it is done is the fact that the geological column and approximate ages of all the fossil-bearing strata were all worked out long before anyone ever heard or thought about radioactive dating."[19] "Not even uranium dating is capable of experimental verification, since no one could actually watch uranium decaying for millions of years to see what happens."[20] So scientific creationism teaches that the fact that early estimates of the age of rocks are supported by a more accurate later technique somehow means that the later technique is wrong!

To account for the earth's geological features by catastrophes like a Biblical deluge is an eighteenth-century idea that in the light of modern geology can only be termed preposterous. The evidence of the fossil record and of erosional features independently indicate, for example, that the Appalachian Mountains are hundreds of millions of years older than the Rocky Mountains. The fossil record shows a progression of life forms that could not conceivably have been buried in so orderly a sequence by a single global catastrophe. Primitive (prokaryotic) cells are in deposits dated at more than a billion years older than the first eukaryotic cells, which are virtually the same in size and shape but enormously more advanced in structure; hardly the sort of thing that could happen in the midst of a worldwide catastrophe.

2. The first law of thermodynamics holds that the sum total of energy in the universe is constant and neither increases nor decreases; the second law of thermodynamics holds that in a closed system, energy tends to go from organized states to disorganization in the form of heat. Creationists take these laws of physics to mean that organized living systems could not have evolved from less organized matter, and that complex organisms could not evolve from simpler ones: "For the evolution of a more advanced organism, however, energy must somehow be gained, order must be increased, and information added. The Second Law says this will not happen in any natural process unless external factors enter to make it happen."[21]

But order arises from disorder all around us. A human body arises from the relative formlessness of a fertilized egg; disordered water molecules form ordered ice crystals in our refrigerators. The reason, of course, is that neither an organism nor anything else except the universe as a whole is a closed system: the earth and its organisms are open systems that acquire energy from the sun to build complexity from simple precursors. As Isaac Asimov has said, the creationist argument from the second law is "an argument based on kindergarten terms [that] is suitable only for kindergartens."[22]

Organisms are programmed by the information in their DNA to synthesize complex molecules with the help of the sun's energy. The information changes by mutation, which if unopposed would break down the order in a living system. But it is not unopposed. Natural selection maintains order and preserves occasional mutations that increase organization while eliminating those that decrease it. Suppose an amoebalike organism engulfs an algal cell and, instead of digesting it, harbors it within its body and uses some of the molecules the alga produces by photosynthesis. A symbiotic combination like this could well survive better than either organism separately. No violation of thermodynamics is involved in this process, which many biologists believe was the origin of the more complex (eukaryotic) plants. Natural selection is the great opponent of the thermodynamic tendency toward disorder.

3. Creationists insist that complex systems could not have arisen by chance, and so must have been formed by an intelligent designer who could impose order on random disorder. They engage in lengthy pseudomathematical "demonstrations" that it is almost infinitely improbable that a particular series of nucleotides should have arisen by chance to form the first nucleic acid. The probability, says Gish, that the molecules in a bacterial cell could be drawn at random is less than one out of a hundred billion.[23] Similarly, if mutations are random, it is infinitely improbable that a mammal should evolve from one-celled protozoan ancestors.

As we have seen, though, there are at least two profound errors in such calculations. One is a failure to recognize that if any one of a great many "successful" events is possible, then the total probability of a successful event is quite high, even if any particular event is highly unlikely. No one said that the first RNA molecule—the first "life"—had to have any particular sequence: any of thousands of different sequences could have formed and then proceeded to replicate themselves, just as they do in laboratory experiments.

Gish says that "origin of life" experiments, in which proteins and nucleic acids are formed spontaneously, are irrelevant, because in a real ocean "efficient methods of producing these compounds would have had to exist, then, since many billions of tons of each would have been required to give a significant concentration in such a vast body of water."[24] But the ocean isn't a big soup tureen in which free nucleotides have to find one another in the vasty deep. Natural environments are full of "traps" for organic molecules—the surfaces of particles, to which organic molecules adhere and on which chemical reactions are catalyzed (see Chapter 7). Moreover, once the first nucleic acid molecules were formed, they didn't change into new, more complex sequences by chance. The ones that were more efficient in capturing organic molecules and replicating themselves more rapidly replaced the less efficient ones by natural selection.

4. Creationists, however, deny that mutation, recombination, and natural selection can form new, complex features. Duane Gish says that

> most mutations result in a change in only one of the several hundred or several thousand subunits in a gene. The change is usually so subtle that it cannot be directly detected by present chemical techniques. The effect on the plant or animal almost always is very drastic, however. Frequently, a mutation proves to be lethal, and it is almost universally, or is universally, harmful. The mutations we see occurring spontaneously in nature or that can be induced in the laboratory always prove to be harmful.[25]

It is not true, however, that mutations are almost universally harmful. Whether mutations that alter the metabolic abilities of bacteria, confer insecticide resistance on a fly, or change the height and growth form of a plant are harmful or beneficial depends on the environment. Evolutionary theory does not postulate that "mutations must be primarily beneficial"—only that some are. Put a culture of bacteria, fungi, or flies into a novel environment and within a few generations it will have evolved improved adaptation, even if, as is easily done with these organisms, you begin with a population of genetically identical individuals, and even if the majority of the mutations in that population are negative.

Creationists do not believe that natural selection can shape mutations into new features:

> Natural selection . . . cannot produce any real novelties. It is a passive thing, a sort of sieve, through which pass only the variants which fit the environment. Those which do not fit are stopped and discarded by the sieving process. However, it can only act on variants which come to it via the genetic potentialities implicit in the DNA structure for its particular kind; it cannot generate anything new itself. The reshuffling, or recombination, of characters already implicitly present in the germ cell certainly does not create anything new in the evolutionary sense.[26]

But if natural selection can preserve individual favorable mutations, as it does, it can also preserve combinations of mutations that are jointly advantageous. Creationists do not deny that the single gene that causes black coloration in the moth *Biston betularia* increased in frequency because dark moths suffer less predation than lighter ones. Now, it so happens that other genes exist in this species that enhance the dark coloration caused by the "major" gene locus, and these genes also have increased in this species over the last century.[27] The combination of several of these genes for dark coloration is more adaptive than any one of them alone.

Another example is provided by an African swallowtail butterfly, in which a gene determines whether a certain part of the wing is white or reddish brown and other genes, closely situated on the same chromosome, determine the black and white pattern of the rest of the wing. Butterflies with certain color patterns, produced by particular combinations of genes, have an advantage because they resemble one or another of several distasteful species of butterflies, and so are avoided by wary predators. Other color combinations are disadvantageous because they don't look like any distasteful species. These disadvantageous combinations, however, occur very rarely in nature, because natural selection— survival of some combinations over others—holds the genes together into the "right" combinations.[28] The point, then, is that if a mutation for reddish coloration and a mutation for a particular spot pattern crop up in different butterflies, genetic exchange during sexual reproduction can bring these mutations together into a new, adaptive pattern, and such a new characteristic can increase in the population because of its enhancing effect on survival. The variation produced originally by mutation, and then mixed into new combinations during reproduction, is preserved by natural selection and molded by it into new adaptations.

The crux of the creationist objection, though, lies in the emphasis on "real novelties." Creationists have responded to the fact that biologists have observed genetic change in organisms by inventing the idea that each "kind" was created with a great variety of genes. However, "Modern molecular biology, with its penetrating insight into the remarkable genetic code, has further confirmed that normal variations operate only within the range specified by the DNA for the particular type of organism, so that no truly novel characteristics, producing higher degrees of order or complexity, can appear."[29]

Modern molecular biology has confirmed no such thing. It has confirmed that mutations can affect a small or a large part of a gene or of a chromosome; that new genetic information can come into existence by the duplication of preexisting genes and by exchanges of nucleotides to form entirely new gene sequences; that mutations can alter the organism's biochemistry a great deal or not at all. Together with the study of development, molecular genetics has shown that even slight genetic changes can provide enzymes with new biochemical functions; can alter the size, shape, and growth rate of every feature of an organism's body; and can produce changes much like those that distinguish different, related, species. The "range specified by the DNA for the particular type of organism" is a creationist fiction for which none of molecular biology offers support.

The "higher degrees of order or complexity" that creationists believe cannot evolve are actually impossible to define. Begin with a reptile, for example, and imagine one of the lower jaw bones becoming larger and the others smaller, so that they finally are disconnected. Is this an increase in complexity? It is one of the chief defining features of the class Mammalia. The single-cusp tooth of the reptiles develops small accessory cusps. Is this a higher degree of complexity? The different variations on the multicusp theme are the basis of much of the adaptive radiation of the mammals into different ways of life, and genetic variations in tooth form are common within many species of mammals. Imagine slight variations in the position of the eyes, from the side of the head toward the front. Such variations in shape and orientation are characteristic of almost every feature of organisms, although this particular one is a major adaptive feature of the primates. Is it really more complex than similar ones in "lesser" species? The "higher degrees of order and complexity" that so impress the creationists are, in a sense, illusory. The "complexity" of a horse or a dandelion is actually just a collection of individual features. Each of these can (and usually did) evolve independently, and each is a not very drastic remodeling of the features of the ancestor. And the material for remodeling is evident in the variation within species.

The creationists continue to argue that variation cannot transcend the limits of the "kind"—a Biblical term that has no meaning in modern taxonomy. Yet they have no idea how to define or recognize a "kind." Are lizards and snakes different "kinds" because iguanas are so different from cobras, or the same "kind" because there are so many intermediate snakelike lizards and lizardlike snakes? This vagueness is convenient for the creationist argument, of course, because whenever a biologist or paleontologist finds an intermediate between two "kinds," the creationist can claim that they are the same "kind" after all. The argument that genetic changes cannot bring about new, more complex "kinds" of organisms rests on the belief that organisms fall into discrete, higher and lower "kinds." But they do not.

5. Perhaps the favorite creationist theme is the gaps in the fossil record. According to Gish, the rich fossil collections in museums ought to contain thousands of transitional forms, yet the most eminent paleontologists agree that the origins of most major animal groups are not

revealed by the fossil record. "As a matter of fact," Gish says, "the discovery of only five or six of the transitional forms scattered through time would be sufficient to document evolution."[30]

Consider several specific examples of the gaps that offer Gish and his colleagues such solace: "The oldest rocks in which indisputable metazoan fossils are found are those of the Cambrian period. . . . These animals were so highly complex it is conservatively estimated that they would have required 1½ billion years to evolve. What do we find in rocks older than the Cambrian? *Not a single, indisputable, metazoan fossil has ever been found* in Precambrian rocks!"[31] But this is not true. As paleontologist Preston Cloud says, "Since 1954, a variety of primitive microorganisms have been found to occur through a long sequence of rocks dating back to more than 2 billion years ago. We now also have evidence that a limited variety of multicellular animal life began about 680 million years ago, perhaps 80 million years before shell fossils of the Cambrian, and that higher forms appeared sequentially up to, through, and beyond the Cambrian."[32] Moreover, it is impossible to tell from Gish's text who "conservatively estimated" that the first Cambrian animals would have required 1½ billion years to evolve.

One of the best examples of a fossil intermediate between major groups is *Archaeopteryx*, the first "bird." As far as Gish is concerned, "The so-called intermediate is no real intermediate at all because, as paleontologists acknowledge, *Archaeopteryx* was a true bird—it had wings, it was completely feathered, it *flew*. . . . It was not a half-way bird, it *was* a bird." Gish dismisses the fossil's reptilian features: it had reptilelike teeth, but "while modern birds do not possess teeth, some ancient birds possessed teeth, while some others did not. Does the possession of teeth denote a reptilian ancestry for birds, or does it simply prove that some ancient birds had teeth while others did not?"[33] *Archaeopteryx* had claws on its wings—but so does a modern bird, the hoatzin. So as far as Gish is concerned, clawed wings say nothing about ancestry.

There is no reason, of course, why ancestral characteristics shouldn't persist in some descendant species and be lost in others. *Archaeopteryx* isn't an intermediate between reptiles and birds merely because it has teeth and claws. It is an intermediate because it occurs before any of the birds that have more "advanced" characteristics; because it has exactly the characteristics that the ancestors of the birds must have had if they descended from reptiles; because it occurs at the same geological time as the small theropod dinosaurs; and because it is almost identical to these dinosaurs in virtually every characteristic except its feathers. It has a long series of tail vertebrae, unfused back vertebrae, unfused limb bones, a rudimentary series of breastbones, solid limb bones, and innumerable other characteristics that are indistinguishable from other small reptiles of the time (see Figures 10 and 11, Chapter 4). It is classified as a bird because of only one characteristic—feathers. Use any other criterion, and it will be classified as a reptile. Organisms, especially extinct ones, don't fall nicely into categories.

Gish hastens past the reptile-to-mammal transition with very little comment, presumably because paleontologists find the distinction between mammal-like reptiles and reptilelike mammals to be totally arbitrary. The usual criterion is that a fossil is classified as a reptile if its lower jaw has several bones, of which the articular bone connects to the quadrate bone of the skull. If the lower jaw consists only of the dentary bone, connecting to the squamosal bone of the skull, it is arbitrarily classified as a mammal. The quadrate and articular bones of the reptile form two of the three auditory ossicles of the mammal's middle ear. According to Gish, these differences "have never been bridged by intermediate series. . . . There are no transitional forms showing, for instance, two or three jaw bones, or two ear bones."[34]

There are, of course, other differences between undisputed reptiles and undisputed mammals: mammals have two occipital condyles; a secondary palate; legs held under the body; teeth modified into incisors, canines, and molars. Reptiles have a single occipital condyle and usually lack a secondary palate; usually all the teeth are single cusps; and the legs are usually splayed out to the side. In reptiles, an opening on the side of the skull is separated from the eye socket by a bony bar; in mammals the bar is absent.

In the synapsid reptiles of the Permian and Triassic times, the bony temporal bar showed a trend toward reduction, and the teeth were somewhat different in different parts of the jaw. Paleontologist Edwin Colbert describes one group of them, the therapsids, as follows:

> The quadrate and quadratojugal bones were reduced to very small elements, often loosely connected to the skull, as contrasted with the large quadrate in most reptiles. In the more advanced therapsids there was a secondary palate below the original reptilian palate. . . . In the lower jaw the dentary bone tended to enlarge at the expanse of the other jaw bones. . . . The differentiation of the teeth progressed in the therapsids to high levels of development, with the advanced genera showing sharply contrasted incisors, canines, and cheek teeth. In many therapsids the occipital condyle became double, as in the mammals.[35]

Also, the legs were held more or less under the body, permitting more efficient running.

The most mammal-like of the therapsids were the theriodonts, such as the wolflike *Cynognathus* of the lower Triassic. It had small incisors, a greatly enlarged canine tooth, and cheek teeth with accessory cusps, just like many of the later mammals. The secondary palate was well developed, and there was no bony bar behind the eye. In almost all respects its skeleton was mammalian, as Colbert remarks, except for the lower jaw: "The dentary bone was so large as to form almost the entire body of the lower jaw, with the bones behind the dentary quite small and crowded." The quadrate and articular bones were very small.

From the therapsids, one moves imperceptibly in the Triassic and lower Jurassic to the ictidosaurs, in which, according to Colbert, many of the therapsid characteristics "were carried even farther toward the mammalian condition. . . . However, the ictidosaurians retained the quadrate bone in the skull and the articular bone in the lower jaw, even though these elements were reduced to such a small size that they were almost no longer functional. For this reason and perhaps no other the ictidosaurs are classified as reptiles. . . . All of which indicates how academic is the question of where the reptiles leave off and the mammals begin."

Of course, if one defines a mammal as a creature with one jaw bone, and defines a reptile as a creature with more than one, then all fossils will be either mammals or reptiles. There can be no intermediates by such a definition. But if the ictidosaurs, retaining only a hint of the reptilian jaw structure amid a panoply of mammalian characteristics, aren't intermediates, what would be?

Contrary to creationist claims, the transitions among vertebrate classes are almost all documented to a greater or lesser extent. *Archaeopteryx* is an exquisite link between reptiles and birds; the therapsids provide an abundance of evidence for the transition from reptiles to mammals. Moreover, there are exquisite fossil links between the amphibians and reptiles (the seymouriamorphs) and between the crossopterygian fishes and the amphibians (the icthyostegids). Of course, many other ancestor-descendant series also exist in the fossil record. I have mentioned (Chapter 4) the bactritid-ammonoid transition, the derivation of several mammalian orders from condylarthlike mammals, the evolution of the horses, and of course the hominids.

Undeniably, the fossil record has provided disappointingly few gradual series. The origins of many groups are still not documented at all. But in view of the rapid pace that evolution can take, and the extreme incompleteness of fossil deposits, we are fortunate to have as many transitions as we do. The creationist argument that if evolution were true we should have an abundance of intermediate fossils is built by exaggerating the richness of paleontological collections, by denying the transitional series that exist, and by distorting, or misunderstanding, the genetical theory of evolution.

In a typical sally, Gish says rodents are so abundant and rich in species that we should certainly find fossil evidence of their origin.[36] However, the fact is that these animals are so small and fragile that few of them have been preserved as fossils. In addition, the supposition that evolution proceeds very slowly and gradually, and so should leave thousands of fossil intermediates of any species in its wake, has not been part of evolutionary theory for more than thirty years. Since we now know that populations harbor an enormous amount of genetic variation, we are aware that evolution does not have to wait for rare favorable mutations. When environments change, that genetic variation is quickly mobilized, and a species evolves to a new genetic equilibrium. G. Ledyard Stebbins and Francisco Ayala point out that laboratory populations of fruit flies have increased 10 percent in body size within twelve years; if human brain size had evolved at the same rate, the increase in cranial capacity from *Homo erectus* to *Homo sapiens* could have taken less than 13,500 years—"a geological instant."[37] It is hardly surprising that the fossil record, usually based on a stratum here and a stratum there, separated by a few million years, gives the appearance of "jumps" from one creature to another. It is usually too coarse to give a more detailed picture.

6. One of the oldest objections to Darwin's notion of gradual evolutionary change is that complex organs would have no survival value when they are just beginning to develop. The classic example is the vertebrate eye, with its complex organizationof retina, lens, focusing muscles, iris, and cornea. How could it have functioned unless all the pieces were in place? Thus, the creationists have said:

> Even if variation, or recombination, really could produce something truly novel, for natural selection to act on, this novelty would almost certainly be quickly eliminated. A new structural or organic feature which would confer a real advantage in the struggle for existence—say a wing, for a previously earth-bound animal, or an eye, for a hitherto sightless animal—would be

useless or even harmful until fully developed. There would be no reason at all for natural selection to favor an incipient wing or incipient eye or any other incipient feature.[38]

Now, there is no genetic reason why all characteristics should evolve by very slight successive modifications. Single mutations that appreciably alter a structure could sometimes be adaptive. This is, indeed, one of the major points of discussion in modern evolutionary biology. Moreover, a structure evolved for one function may be ready and available to serve a new function, so that it doesn't have to develop *de novo*. When the several bones that the reptiles had used for the articulation of the jaw became small and lost their original function, they were in the right place to be modified for sound transmission in the mammalian middle ear.

But the idea that a complex structure must be fully constituted to be functional is simply not true, as the eyes of animals in fact demonstrate. Even in protozoans, there are light-sensitive organelles associated with the cilia (the hairlike structures used for swimming), which permit the organism to orient toward light. In many jellyfishes, worms, and other invertebrates, ciliated cells, in which the light-sensitive structure resembles that of the protozoa, are aggregated into a flat "eye spot." In the worm *Eunice*, light-sensitive cells are covered by a clear cuticle that is slightly thickened, like a flat lens. In another worm, *Nereis*, the light-sensitive cells form a cup, or retina, surmounted by a lens-shaped covering that is almost spherical. Each of these successively more complex eyes is clearly functional, and each slight modification, from an organ that can merely distinguish light from dark to one that can form a progressively more defined image, would be adaptive.[39]

7. In recent years, it has become evident that to attribute all of evolution solely to the joint action of mutation and natural selection is to hold too narrow a view of the evolutionary process. As Sewall Wright and many other evolutionists have long maintained, some evolution occurs by genetic drift, so that neutral, nonadaptive changes can take place. It also seems possible that selection may act at the level of the whole population or species, so that evolution may have to be understood in terms of the replacement of some species by others, in addition to the time-honored mechanism of natural selection, the replacement of inferior by superior genotypes *within* populations. Some evolutionists, especially paleontologists and morphologists, have therefore asserted vigorously that the neo-Darwinian theory, narrowly construed as muta-

tion plus selection within populations, must be expanded into a more pluralistic view that emphasizes these other factors as well.

The creationists have gleefully claimed[40] that these sentiments reflect a "growing repudiation of neo-Darwinian orthodoxy," a rejection of natural selection as a major evolutionary force. For example, Gish interprets the neutralist theory of change by genetic drift this way: " 'Neutralists,' as advocates of this theory are called, believe that many repetitions of this process, along with continued mutations to replenish the gene pool, have been responsible for the evolutionary origin of all living things."[41] But nothing could be further from the truth. Biologists do not hold Gish's all-or-none, black-or-white view of the world. It is very likely that a great deal of genetic change consists of the random replacement of one protein variant by another that is neither better nor worse. But this does not rule out natural selection as the driving force behind new adaptations. Not even the most vehement "neutralist" would argue that chance, rather than natural selection, was responsible for the evolution of eyes, wings, or flowers.

The hypothesis of punctuated equilibrium holds that natural selection generally causes new adaptations to arise only when a new species branches off from its ancestor, which persists unchanged. A corollary of this hypothesis is that a single species doesn't change continually through long periods of time; instead, an evolutionary trend in one direction would be due to a changing succession of species, if, say, large-bodied species tend to escape extinction longer than small-bodied species. One of the proponents of this idea, paleontologist Steven Stanley, notes that there is a random element in this process, because the barriers that isolate populations and permit them to become new species are unpredictable. He goes on to say that "natural selection, long viewed as the process guiding evolutionary change, cannot play a significant role in determining the overall course of evolution,"[42] because the long-term trend is caused by extinction versus survival of whole species. This is a different process from the death versus survival of different genotypes within a single species, which the term "natural selection" embraces.

According to Duane Gish, "Stanley believes that evolution has occurred by abrupt, random production of new species. He offers no explanation whatsoever of how a species may abruptly, at random, produce new species."[43] But that is not so. Stanley devotes a good deal of space to Mayr's theory of how natural selection can rapidly transform small, localized populations into new, adaptively different species. Gish, however, goes even further, and sees in this theory the death knell of

Darwinism: "If what Stanley says is true, the theory of evolution, devoid of any real evidence from the fossil record to support it, is devoid even of a theoretical framework."

In reality, what Stanley is doing is enriching the theoretical framework of evolution by adding the role of species extinction to that of natural selection. He is suggesting that if the environments to which species become finely adapted by natural selection change drastically, only certain species will survive to determine the subsequent course of

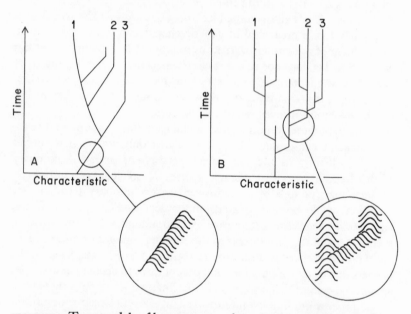

FIGURE 27. Two models of how a group of species (1, 2, 3) may arise from a common ancestor. The lines represent the average state of an evolving characteristic; the inserts show that there is variation around this average at all times (cf. Figure 2). The "gradualistic" model (A) shows lineages changing gradually over a long period of time and then reaching an equilibrium. The model of punctuated equilibrium (B) proposes that most of evolution occurs during short periods of time when new species branch off from ancestral lineages that persist with little or no change. To a large degree, the differences between the two models are a matter of time scale. Compare this diagram with the neo-Darwinian model of evolution illustrated in Figure 2 (Chapter 2).

evolution. Stanley finds in natural selection the mechanism that adapts individual species, in the short term, to their individual environments; but he hypothesizes that the overall history of life, the "overall course of evolution," depends on the long-term succession of environmental events that determine which species will die out and which will thrive. His hypothesis may or may not be valid, but it is a plausible, testable idea,[44] and the controversy it has generated is the mark of an active, healthy science.

8. Finally, creationists claim that evolution is a "religion" that requires faith, since, according to them, it operates too slowly to be observed, cannot be proven true, and cannot be tested—and so, they say, cannot be a scientific hypothesis. In fact, of course, evolutionary change can be observed; and even if it couldn't, it can be deduced from the characteristics of fossil and living organisms. The theory certainly can be tested, and has been, innumerable times. It cannot be proven absolutely true, but as I pointed out in Chapter 9, neither can any other scientific theory. The idea of evolution, like the theory of the solar system or the theory of chemical bonds, just has so much evidence in its favor that it is almost inconceivable that another scientific theory could take its place.

I s it possible, despite all the evidence, that evolutionary scientists are wrong? Here is what Duane Gish has to say on the subject:

> Why have most scientists accepted the theory of evolution? Is the evidence really that convincing? This seems to be the clear implication. On the other hand, is it possible for that many scientists to be wrong? The answer is an emphatic "YES!" Consider for a moment some historical examples. For centuries the accepted scientific view was that all planets revolved around the earth. This was the Ptolemaic geocentric theory of the universe. Only after a prolonged and bitter controversy did the efforts of Copernicus, Galileo, and others succeed in convincing the scientific world that the Ptolemaic system was wrong and that Copernicus was right in his contention that the planets in the solar system revolved around the sun.[45]

What an extraordinary reinterpretation of history! The geocentric theory of the universe was a theological doctrine developed by St. Clement of Alexandria, Dionysius the Areopagite, the twelfth-century

theologian Peter Lombard, and St. Thomas Aquinas. For fear of ecclesiastical censure, Copernicus did not dare publish his theory for more
than thirty years, until, in 1543, he lay dying. His fear was justified;
Giordano Bruno was burned alive in 1601 for his heretical views on
astronomy. When Galileo affirmed that the Copernican theory was true,
he was forbidden by the Pope and the Roman Curia to teach the theory,
and "all books which affirm the motion of the earth" were forbidden.
Professors throughout Europe were convinced that the theory was true
but were forbidden to teach it. As Galileo revealed more and more
forbidden astronomical phenomena, a Father Caccini preached a ser-
mon from the text "Ye men of Galilee, why stand ye gazing up into
heaven?"; he declared that "geometry is of the devil" and that "mathematicians should be banished as the authors of all heresies," and was
elevated in the Church hierarchy. Galileo was forced by the Inquisition
to acquiesce in the injunction "In the name of His Holiness the Pope
and the whole Congregation of the Holy Office, to relinquish altogether
the opinion that the sun is the centre of the world and immovable, and
that the earth moves, nor henceforth to hold, teach, or defend it any way
whatsoever, verbally or in writing." Finally, for writing an imaginative
account of the Copernican theory, he was imprisoned, threatened by the
Pope with torture, and forced to kneel in public and recant his heresy:
"I, Galileo, being in my seventieth year, being a prisoner and on my
knees, and before your Eminences, having before my eyes the Holy
Gospel, which I touch with my hands, abjure, curse, and detest the error
and the heresy of the movement of the earth."

However difficult it may sometimes be, the world of science can
relinquish old theories when evidence favors new ones. But there was
no "scientific world" until Copernicus and Galileo; there was only a
theological world as determined then to conceal and deny the truth as
fundamentalist creationists are today. The geocentric theory was unproven; it was untestable; it was held only because it fit Scripture.
Sooner or later it had to yield to a scientific theory, a theory that made
verifiable predictions. Andrew Dickson White, in *A History of the Warfare of Science with Theology in Christendom*, [46] tells us that the opponents
of Copernicus protested to him that, if his theory were true, Venus, like
the moon, would show phases in the course of its revolution about the
sun. Copernicus answered, " 'You are right; I know not what to say; but
God is good, and will in time find an answer to this objection.' The
God-given answer came when, in 1611, the rude telescope of Galileo
showed the phases of Venus." If creationism is science, let it make a
single prediction that could show it right or prove it wrong.

ELEVEN

THE CASE
FOR EVOLUTION

L ife is too short to occupy oneself with the slay-
ing of the slain more than once.
—T. H. HUXLEY, 1861

Creation and evolution, between them, exhaust the possible explanations for the origin of living things. Organisms either appeared on the
earth fully developed or they did not. If they did not, they must have
developed from preexisting species by some process of modification. If
they did appear in a fully developed state, they must indeed have been
created by some omnipotent intelligence, for no natural process could
possibly form inanimate molecules into an elephant or a redwood tree
in one step. If species were created out of nothing in their present form,
they will bear within them no evidence of a former history; if they are
the result of historical development, any evidence of history is evidence
of evolution.

If species are the products of creation, what should we expect to see?
According to the creationists, "The First Cause of all things must be an
infinite, eternal, omnipotent, omnipresent, omniscient, moral, spiritual,
volitional, truthful, loving, living Being!"[1] That is, the nature of the
Creator is inferred from the creation. But this argument from design is
a two-edged sword. Gerard Manley Hopkins may find that "the world

is charged with the grandeur of God," but Shakespeare can just as well say in *King Lear* that "as flies to wanton boys are we to the gods; they kill us for their sport."

If we find the natural world to be full of useless features, inadequate design, shoddy workmanship, and harshness or cruelty, we must either conclude that the personal Creator envisioned by creationists is cruel, thoughtless, and incompetent, or else conclude that He is indeed omnipotent, omniscient, and loving, but also capricious and arbitrary. But if the Creator acts at whim, without consistency or reason, we can make no predictions whatever: species may be well adapted or not, have useful or useless organs, and we are left ascribing them to the inscrutable acts of an inscrutable God. We are left without hope of explanation or understanding; and science, which searches for understanding by making and testing predictions, ceases to exist.

The case for evolution then has two sides: positive evidence—that evolution has occurred; and negative evidence—that the natural world does not conform to our expectation of what an omnipotent, omniscient, truthful Creator would have created. If the creationist replies that everything in the world, no matter how arbitrary, useless, or cruel, is just what we should expect of the Creator's infinitely inscrutable wisdom, he is playing Dr. Pangloss to our credulous Candide; and he is tacitly admitting that creationism can predict nothing, and so cannot be science.

We look at the design of organisms, then, for evidence of the Creator's infinite intelligence, and what do we see? A multitude of exquisite adaptations, to be sure: the bones of a swallow beautifully adapted for flight; the eyes of a cat magnificently shaped for seeing in the twilight. But if we look further, we find that the bones of the flightless dodo and penguin are also hollow, as if adapted for flight; and that the mole and the cave salamander also have a lens and retina that serve no function. Every organism has such vestiges of structures that can only be the useless remnants of past adaptations. Why should we have wisdom teeth, unless our jaws have become shorter, so that our ancestors' teeth no longer fit? Why should we, like other primates but unlike almost all other vertebrates, require vitamin C in our diet, unless we stem from ancestors who got enough vitamin C in their diet of fruit? Do we find here evidence of wise design?

Look further in the living world and you find animals and plants that make do with inferior adaptations. Primates manipulate their food with a dexterous opposable thumb, but the giant panda makes do with a clumsy thumblike structure modified not from the first finger but from a wrist bone. Consider how immensely useful photosynthesis would be to animals when food is in short supply. Some marine animals such as corals indeed make use of photosynthesis by harboring algae in their bodies, but no higher animals have been endowed with their own photosynthetic mechanism. Compound eyes are one of the major adaptations of adult insects such as butterflies and bees; yet no caterpillar in the world has them, however useful they might be. Is it easier to believe that an omnipotent Creator withheld adaptations that would make life easier, or to believe that this is not the best of all possible worlds, and that species make do with the best genetic variations that happen to become available?

When we compare the anatomies of various plants or animals, we find similarities and differences where we should least expect a Creator to have supplied them. Is it not strange that a Creator should have endowed bats, birds, and pterodactyls with wings made out of the same bony elements that moles use for digging and penguins use for swimming? Is it not stranger still that instead of modifying these bones for flight in the same way, the Creator should have decreed that the bat's wing be made by lengthening four fingers, the pterodactyl's by lengthening only one finger, and the bird's by shortening the hand and equipping it with feathers? If evolution happens, we should expect different organisms to evolve different solutions, by chance, to similar problems; but an omniscient Creator shouldn't have to experiment with different designs. Birds and mammals, being warm-blooded, have only one aortic arch instead of two as in amphibians and reptiles, for more efficient transport of blood from the heart; but can any creationist explain why birds retain the right aortic arch while mammals retain the left one?

Embryology, too, reveals that related species are similar in ways that make no adaptive sense. The embryos of whales and anteaters develop teeth and then absorb them before birth. This makes sense if they carry in their genes the imprint of their history; but could any creationist have predicted these embryonic patterns by the argument from design? Or is the Creator trying to trick us into believing in evolution?

Molecular biology finds at every turn patterns that make sense in an evolutionary interpretation, and no sense in a creationist view of the world. The similarity of DNA is greatest not between animals that have similar adaptations and similar modes of life, but between species that on anatomical grounds are believed to be evolutionary relatives. The DNA of parasitic worms is similar not to that of other parasites but to the DNA of their nonparasitic relatives. Can this be part of a Creator's design? Can we attribute to an omniscient Creator the fact that bacteria have "silent" genes that are never expressed and appear to have no function?

The modern theory of evolution says that organisms should experience random mutations that are not designed to be adaptive but may turn out fortuitously to be so; and that adaptation proceeds by natural selection—the success of the fitter at the expense of the less fit. The creationists admit that species can undergo limited adaptive changes by the mechanism of mutation plus natural selection. But surely an omniscient and omnipotent Creator could devise a more foolproof method than random mutation to enable his creatures to adapt. Yet mutations do occur, and we have experimental demonstration that they are not oriented in the direction of better adaptedness. How could a wise Creator, in fact, allow mutations to happen at all, since they are so often degenerative instead of uplifting? According to the creationists,[2] there is "a basic principle of disintegration now at work in nature" that we must suppose includes mutation. But why should the Creator have established such a principle? Didn't He like the perfection of His original creation?

And natural selection: what cruelty and waste it seems to entail! As the creationists say, surely the God of the Bible could not have invented such a mechanism to build adaptations and maintain them. But however distasteful it may appear, natural selection is found operating in every species that has ever been studied. It would be hard to imagine a crueler instance of natural selection than human sickle-cell anemia—part of the population being protected against malaria at the expense of hundreds of thousands of people condemned to die because they have inherited a disastrous gene that happens to be worse for the malarial organism than for some of the people who carry it. And such examples are not limited to our sinful species; do not suppose that sickle-cell anemia is part of the fall from grace. In the forests of South America there is a species of fruit fly in which exactly the same principle remorselessly kills many of the offspring of every fly: heterozygotes for a particular gene survive, but many of their homozygous offspring die in their infancy.[3]

And what are we to make of those cases of natural selection run amok? Are they instances of the divine wisdom? Does it accord with the divine sense of harmony that male elephant seals should battle so furiously for females that great numbers of them die of bloody wounds? Did God require the peacock to carry such long feathers that it can hardly fly, just so that it could attract females and insure the propagation of the species?

Looking to ecology, we should expect balance and harmony in a perfect creation. But the "balance of nature" is a myth created by visionary Victorians and perpetrated upon us by the solemn voice-overs on television natural history shows. Yes, there is a nitrogen cycle, whereby various bacteria transform atmospheric nitrogen into compounds that plants and animals can use. But such bacteria are far from ubiquitous, and there are many places on earth where usable nitrogen and other essential elements are in very short supply. True, predators sometimes prevent population explosions of their prey. But very often they do not. Even in natural ecosystems untouched by human disturbance, plague locusts and other species erupt in vast numbers, lay waste the land, and perish by the millions. Unless prey species have adequate defenses, their predators and parasites do indeed extinguish them. The fungus that extinguished the chestnut and the rats that have brought many species of island birds to extinction have not been stayed by the divine desire for natural harmony. A wish for harmonious coadaptation should have impelled the Creator to endow species with the ability to check their own increase and stop short of overpopulation, but species do not have this ability. If species do stay within certain limits of population size, it is only because of the forces of nature that kill thousands for every one that survives.

And why should there be more than a million species of animals and more than half a million of plants? Creationists know why: everything was created to serve a purpose, and that purpose is profoundly anthropocentric. "The earth was created specifically to serve as man's home," and "all . . . created systems must in some way be oriented man-ward, as far as purposes are concerned."[4] They would have us believe that the 250,000 known species of beetles, and untold numbers of undescribed species, exist to serve humanity; that Antarctic birds possess unique species of lice for the benefit of man; that the lung worms that infest snakes and the schistosome worms that kill hundreds of thousands of people each year are part of God's gift to the human species. Hundreds of thousands of species of ammonites, brachiopods, graptolites, alcyonarians, and other extinct invertebrates, of which not one person in a million has ever heard, existed and died out millions of years before humans walked the earth; was their "purpose" only to make us puzzle over their remnants in the rocks, deluding us into believing in evolution?

Finally, the fossil record can never be fitted into a creationist interpretation. Nowhere does the absurdity of their arguments become more evident than in their frantic, fanciful attempts to explain the fossil record and the fact that more than 90 percent of the species the Creator is supposed to have created became extinct, just as if no one cared. At least the pre-Darwinian creationists, in their devoted search for enlightenment, supposed that God might have indulged in dozens or hundreds of successive creations, resulting in the orderly sequence of fossil faunas. But because fundamentalist creationists have to squeeze all of creation into six days, the only way they can explain the geological record is to invoke great cataclysms associated with the Biblical flood. They visualize gigantic currents of water, enough to cover the world, pouring from the skies, accompanied by vast volcanic explosions and massive movements of the continents, great increases in the temperature and turbidity of the oceans, and finally the deposition of sediments as the oceans settled. "The very complexity of the model makes it extremely versatile in its ability to explain a wide diversity of data (although, admittedly, this makes it difficult to test)."[5] That's the understatement of the decade.

Now listen to some of the "obvious predictions" of the flood model.[6] Animals that lived together in the same ecological communities would normally be buried together (despite the great vortices and earth movements?). Marine fishes would be preserved in higher rocks than invertebrates because "they live at higher elevations" (as if fishes and invertebrates did not coexist in every marine community in the world). Amphibians and reptiles would be fossilized in still higher rocks because they are found at the interface between land and water (but how is it that marine shells are found on mountaintops, and why weren't amphibians swept into marine sediments?). Few birds would be found at all, because of their mobility (where did most of the birds go? I don't know any birds that can fly steadily, without food, for forty days and forty nights). Higher animals such as land vertebrates would be segregated vertically in order of complexity, because the "more diversified" animals could escape the cataclysm longer and move to the mountaintops (could mice really move faster than the small, swift dinosaurs? And why didn't the winged pterodactyls make it up to the top?).

As long as we're entertaining such thoughts, let's go a bit further. The creationists' source of truth is Genesis, wherein we find the story of the flood. Genesis also says that God commanded Noah to take into the ark a pair of every living thing that creeps upon the face of the earth, and we are told that Noah obeyed the Lord's command. But if all the millions of extinct species in the fossil record perished in the flood, Noah didn't really obey the Lord, even though the Bible says he did. If he did take them in, the Lord, for unknown reasons, must have let them perish after they left the ark. Nonetheless, at least the million species of animals and presumably the half million or more plants that we have today survived. They must have been in the ark—all 2 million individual animals. Australian kangaroos, South American boa constrictors, Arctic foxes, New Zealand kiwis, and 250,000 species of beetles. Not to mention all their parasites. And, assumedly, food for a million species for a month or so. I suppose all these species lived together in the Middle East, within easy reach of the ark, and that Noah was the best animal collector in the history of the world. Don't forget, of course, the thousands of species of fresh-water fishes. They couldn't have survived in a raging, salty sea, so the ark must have had a big aquarium in it.

Can you believe that any grown man or woman with the slightest knowledge of biology, geology, physics, or any science at all, not to speak of plain and simple common sense, can conceivably believe this? Can you for one moment imagine that this is supposed to be taught to children in the name of science? With or without the story of the ark, the flood cannot conceivably account for the facts of geology and paleontology. Not only are the creationists who propound such nonsense abysmally ignorant of, or blind to, the most elementary facts of biology and geology; not only are they willing to invent stories that defy every law of nature to save their myth of creation; but they have the arrogance to claim that these stories are "science," and that their "science" is just as good as that of thousands of geologists and biologists who have devoted their lives to careful experimentation, observation, and logic. What conception can a young person have of how to seek knowledge if he or she learns that a myth of gigantic earth cataclysms unlike anything known to science, a myth that contains within it the most absurd contradictions and defies every fact of biology, deserves "equal time" with hypotheses that have been tested and supported by countless careful experiments and observations?

On the positive side of the ledger, let me briefly summarize the evidence for evolution. First, the theoretical mechanisms of evolutionary change have been abundantly documented by experiment and observation. Mutations in every kind of characteristic—anatomical, physiological, biochemical, behavioral—are known to occur. Some of these are indeed unconditionally harmful, but some are positively beneficial. Most mutations do not damage the organism irreparably, but merely cause slight changes in a characteristic—the size or shape of an organ, the activity of an enzyme. The genetic variations that arise by mutation accumulate, so that every population contains an immense amount of genetic variability that enables it to change rapidly when environmental conditions are altered.

Natural selection operates when some genetic variants are more successful in surviving and reproducing than others. When the environment isn't changing, selection tends to eliminate genetic variants that deviate from the most favored types in any direction. When the environment changes, formerly inferior variants may become superior and replace the prevalent types. Both forms of natural selection—maintenance of the status quo, and change toward a better-adapted form—have been observed many times in natural populations of plants and animals. Different populations of a species adapt to different environmental conditions. If they acquire differences that prevent them from mating with each other, they become different species. The evolution of such reproductive barriers has also been observed both in experimental situations and in the wild.

Thus we know from direct observation and experiment that the ingredients of evolutionary change are real and potent, just as a geologist knows that erosion is a fact of physical geology. Over the course of millions of years, it is inconceivable that erosion and other observable geological mechanisms should fail to create great gorges and canyons; and it is just as difficult to imagine that mutation and natural selection should fail to create great changes in species over vast periods of time.

The changes that mutation and natural selection can bring about in any one species within the short span of human observation are limited in degree; we can see one species of fly give rise to another, but we do not expect to see flies transformed into fleas in laboratory experiments. That would be asking too much. Such great alterations can only be formed by successive transformations of intermediate steps. If, however, we look at any major group such as the insects, we see a continual gradation of differences from the very slight to the very great. Different families or orders of insects or mammals differ only in degree: they have the same structures, but many of the structures have been greatly modified in size, shape, and arrangement.

But science does not require direct observation to verify its hypotheses; its most powerful technique is the testing of predictions. We cannot observe the orders of mammals diversify from mammal-like reptiles, nor the great diversity of living things flower forth from a single Precambrian ancestor. But we can test the predictions of the hypothesis that living things have a common ancestry.

We predict, first of all, that all living things should share certain common characteristics, and they do. All species, from bacteria to mammals and trees, use the same genetic code, in which the same nucleotide sequences code for the same amino acids. All species use "left-handed" amino acids to make proteins out of. (Amino acids, like other organic molecules, come in two forms that are mirror images of each other but otherwise have the same chemical properties.) From a chemical point of view, this universality is not necessary. Different genetic codes would have served equally well to make proteins out of, and right-handed amino acids could have worked just as well as left-handed ones. The only possible reason for these chemical universalities is that living things got stuck with the first system that worked for them. Once the genetic code was established, no species was ever free to try a new one. A mutation that caused the nucleotide sequence UUU to code for glycine instead of phenylalanine would have messed up all the species' proteins. Similarly, a species whose ancestor used only left-handed amino acids wasn't free to use right-handed ones. They wouldn't have fit properly into the proteins. Where a Creator would have been free to use different biochemical building blocks for different species, evolution was not free: the history of the earliest organisms determined everything that happened thereafter.

The hypothesis of evolution predicts, next, that organisms should share various characteristics in a hierarchical arrangement. As species split and then give rise to more species, they will form a phylogenetic tree. All the members of any major branch of the tree will share characteristics that diverged early in history. Within these major branches, related "twigs" that have branched off more recently will share their own special sets of more newly evolved features.

Allowing for the reversals and parallelisms that do sometimes occur, the world's species do fit this hierarchical arrangement very well. Within the vertebrates, we distinguish major branches such as fishes and tetrapods—the four-legged animals that have a particular arrangement of bones in their limbs. The tetrapods divide into amphibians, which lack an amnion, and the reptiles, birds, and mammals, which possess one. Within the mammals, we distinguish primitive species such as marsupials, which lack a placenta, from the more advanced placental mammals, and so on. The fact that we can use a few characteristics to make a treelike diagram accords with the idea of evolution, but of course it doesn't prove that evolution has occurred. However, the hypothesis of evolution says that there must be only one real evolutionary tree, and if this is so, then it should be possible to find different sources of data that all independently give the same evolutionary diagram. In other words, if structural features indicate that placental mammals diversified after mammals branched off from reptiles, independent lines of evidence should indicate this too.

In fact, there are many independent lines of evidence. One is geological: knowing that the continents have been drifting apart ever since the Permian, we predict that groups which supposedly evolved late in evolutionary history should be more restricted to one or a few continents than groups which supposedly evolved early. This is in fact the case. The fossils of mammal-like reptiles are broadly distributed over all the continents, and so are most of the primitive mammals such as marsupials. But orders and families of mammals such as carnivores and hoofed mammals, which are so similar to each other that they are thought to have diverged more recently, are more restricted in distribution. Many of these groups have never been found, either as fossils or living species, in Australia or South America, which broke away from the other landmasses before groups such as elephants, horses, or apes evolved.

Another independent line of evidence is biochemical. The evolutionary classification of vertebrates was derived before Watson and Crick ever thought of DNA as the genetic material. Since that time it has been possible, either directly by examining the DNA, or indirectly by examining the proteins that the DNA codes for, to measure the genetic similarity of species. Almost invariably these investigations give the same answer that evolutionary taxonomists had found on the basis of morphology: pigs and cows are more similar to each other than they are to dogs; these together are more similar to each other than to primitive mammals such as kangaroos; all mammals are more similar to each other than to reptiles; and so on. Mounting molecular evidence indicates that many of the differences and similarities in DNA sequence are not due to similarities in the organism's structure and function. Rather, much of the variation between species at the level of DNA seems to be in "neutral" changes that have developed by chance (genetic drift): the genes differ somewhat from species to species in their exact nucleotide sequence, but the proteins that the genes make serve the same function. If this is true, divergence in structure has occurred independently of much of the divergence in DNA, so that the two sources of evidence of degree of relationship reinforce one another and are not merely two ways of looking at the same thing.

A major prediction of evolution is that organisms should carry within them the evidence of their history. The evidence of history lies partly in embryology (all vertebrates have similar embryos; terrestrial salamanders go through an "aquatic" larval stage before they hatch from the egg; fetal anteaters have teeth); partly in behavior (our body hairs are erected when we are afraid, just as in the "fight or flight" reaction of more heavily furred mammals); and partly in the useless vestigial structures that every species possesses (there is no functional explanation for the rudimentary eyes of cave animals; the tiny, useless legs of many snakelike lizards; the vestiges of the pelvis in pythons). The effects of history are abundantly evident in the distribution of organisms. There are no native land mammals in Hawaii, not because they cannot survive there, but because they evolved on continents and couldn't cross the Pacific.

Finally, there is the evidence of the fossil record. We predict that if evolution has occurred, old rocks will forever lack fossils of many species that had not yet evolved. We have never found a fossilized mammal or flowering plant in Silurian deposits, and we never will. We should and do find, on the contrary, that the early rocks contain organisms such as lungfishes and cockroaches that are believed to be ancient groups on the basis of entirely independent evidence from the anatomy and biochemistry of their living representatives. Similarly, groups such as elephants and ants which the anatomy of living species tells us are more recently evolved do not appear except in more recent geological formations. On occasion we should and do find gradual transformations from primitive ancestors to modified descendants in exceptionally good fossil deposits. The African record of human evolution is just one of the many cases in which transitional series from ancestors to more modern descendants have come to light.

To pass from primordial molecules to the first cell, and from the first cell to complex animals and plants, had to take time—billions of years. Physicists with their radioactive clocks, astronomers with their "red shifts," geologists with their measurements of continental drift and other earth processes, biologists with their coral clocks—all converge on the same answer: the earth is more than 4 billion years old.

The great age of the earth is a fact. The drift of continents is a fact. It is a fact that the earth has supported different species at different times. It is a fact that species are related to each other by descent. It is a fact that beneficial mutations alter species in every imaginable way. It is a fact that the environment selects some genetic variations for survival and others for extinction. It is a theory that the processes of mutation, recombination, genetic drift, natural selection, and isolation can account for the historical products of evolution, but it is a fact that evolution has occurred. That is the message of the hundred years of biology, geology, physics, and chemistry that have elapsed since Darwin's death.

TWELVE

THE SOCIAL ROLE OF EVOLUTION

> O ur remedies oft in ourselves do lie
> Which we ascribe to heaven: the fated sky
> Gives us free scope; only doth backward pull
> Our slow designs when we ourselves are dull.
> —SHAKESPEARE, *All's Well That Ends Well*

"The growth of a large business is merely a survival of the fittest. The American Beauty rose can be produced in the splendor and fragrance which bring cheer to the beholder only by sacrificing the early buds which grow up around it. This is not an evil tendency in business. It is merely the working-out of a law of nature and a law of God." This encomium to natural selection is not Darwin's, nor that of an atheistic humanist or a free-thinker. It is the voice of John D. Rockefeller.

For at least thirty years, in the late nineteenth and early twentieth centuries, evolution was more American than apple pie. Darwinism, in fact, was more enthusiastically received in America than anywhere else —and not by social reformers or liberals, but by social and economic conservatives. To the average businessman and social theorist, evolution stood for the survival of the fittest, for slow and steady change rather than revolutionary jumps, and for progress. What could be better suited than Darwinism for a country that upheld individual initiative and laissez-faire competition, and made a religion of the concept of progress? What could better suit those who held the reins of power than the concept that slow, progressive change was inherent in natural law?

In reality, though, Social Darwinism, the doctrine that human progress is the outcome of competition and struggle among individuals, races, and nations, had its origin before Darwin. Its chief proponent was Herbert Spencer, who attempted to synthesize all of physics, biology, and human history into a grand evolutionary world view before Darwin published the *Origin of Species*. Spencer was a fierce individualist who believed that competition was the driving force in all of nature. It was he who coined the phrase "the survival of the fittest," which Darwin later used, and held that this process was the source of progress. Developing this idea into a social philosophy, Spencer held that social progress could only occur if the fittest (which he never defined) were allowed free rein. Thus he held an ultraconservative social philosophy that would not sound strange in the mouths of many of today's conservatives: to Spencer, any form of governmental regulation of business, state aid to the poor, state support for education, or even tariffs and governmental postal systems were anathema.

This philosophy endeared Spencer to the giants of American business. Andrew Carnegie became one of Spencer's close friends, and Spencer's speaking tours of the United States were immensely popular. His supporters, such as Yale sociologist William Sumner, used Spencer's and Darwin's writings to argue that "millionaires are the product of natural selection"; that if competition were not given free reign we should be carried backwards by "the survival of the unfittest"; that social change moves according to inexorable natural laws, so that "it is the greatest folly . . . to sit down with a slate and pencil to plan out a new social world." Darwin, who was not a Social Darwinist, did not join the fray, but was distressed. "I have received in a Manchester newspaper," he wrote ruefully to Lyell, "rather a good squib, showing that I have proved 'might is right,' and therefore that Napoleon is right, and every cheating tradesman is also right."

The perversion of biology that was Social Darwinism had more evil uses than the support of laissez-faire capitalism. Social Darwinism went hand in hand with the common belief that all differences among people —in intelligence, drive, criminality—were fixed hereditary traits that could not be altered by the environment. Thus from the turn of the century until about 1920 a strong eugenics movement flourished that advocated social means of encouraging "superior" people to have more children and of discouraging or preventing "inferior" people from breeding. Edward Lee Thorndike, for example, argued that "there is no so certain and economical a way to improve man's environment as to improve his nature." From this, it is but a short step to racism, and the adherents of racial superiority, both in this country and abroad, held that the Anglo-Saxon and Teutonic races were destined by evolution and justified by natural law to hold sway over the "lower" races.

Such racism went hand in hand with imperialism. Listen to the Reverend Josiah Strong, whose 1885 book *Our Country: Its Possible Future and Its Present Crisis* sold 175,000 copies in English alone:

> There is apparently much truth in the belief that the wonderful progress of the United States, as well as the character of the people, are the results of natural selection; for the more energetic, restless, and courageous men from all parts of Europe have emigrated during the last ten or twelve generations to that great country, and have there succeeded best. . . . [When Americans and Europeans move out into the rest of the world,] then will the world enter upon a new stage of its history—*the final competition of races for which the Anglo-Saxon is being schooled.* If I do not read amiss, this powerful race will move down upon Mexico, down upon Central and South America, out upon the islands of the sea, over upon Africa and beyond. And can anyone doubt that the result of this competition of races will be the 'survival of the fittest'? (Emphasis in original.)

But just as the devil can cite Scripture to his purpose, advocates of any political philosophy whatsoever could find something in evolution to support their ideas. One of these was Karl Marx: "Darwin's book is very important and serves me as a basis in natural science for the class struggle in history." Marx was willing to use natural selection as an example of class struggle, but not for individualistic, capitalist competition. A diametrically opposite opinion was voiced by the Russian anarchist Peter Kropotkin in his book *Mutual Aid* in 1902. Kropotkin pointed out that Darwin had explained how cooperation among the members of a species could often be advantageous in their struggle for existence. To Kropotkin, the existence of cooperative social behavior in wolves, monkeys, musk-oxen, and many other animals showed that nature teaches us to eliminate competition by cooperation.[1]

Thus Darwinian evolution has been used in the service of every kind of social philosophy, from the utopian ideals of Kropotkin to the evils of racist imperialism that came to their full expression in Fascism and Nazism. But these philosophies in no way depended on Darwin: they merely used him when he suited their purposes. Racism did not begin in 1859. Gobineau's *Essai sur l'Inégalité des Races Humaines,* a landmark in the history of "Aryanism," appeared in 1853; slavery preceded it by centuries; "might makes right" is perhaps the most ancient of social rules.

Capitalist economics didn't need Darwin. In fact, the idea of natural selection has its roots in the economic theories of Adam Smith and Thomas Malthus, and economists could justly claim that biology had discovered what they already knew. Darwinian biology merely served as a justification for political and social beliefs, by enabling their adherents to claim that their beliefs had the force of natural law.

All these invocations of biology in the service of social theory were, and are, perversions of science both because of their evident scientific errors and their errors of philosophy. The scientific errors are manifold. As a start, it is fallacious to claim that social progress is part of the natural law of evolutionary progress, because "progress" in evolutionary terms is an elusive and probably meaningless concept. Evolution has no predetermined goal. It doesn't even have a direction. To say that the trend of evolution has been toward greater anatomical complexity, or toward greater consciousness as exemplified by the human species, is to ignore the thousands of lineages of plants and animals that have decreased in complexity (parasites, for example), and the thousands of others that have not evolved at all in the direction of greater consciousness. The insects and crustaceans and molluscs are far more "successful" and abundant than the mammals, but they show no movement toward greater complexity or consciousness.

In addition, the "nature red in tooth and claw" version of natural selection on which Social Darwinism is based is a caricature. Natural selection is merely the replacement of the less able by the more able. Able at what? At survival and reproduction. Sometimes there is struggle and competition among individuals and species; sometimes there is not.

Finally, the manifestations of Social Darwinism in the form of eugenics, racism, and imperialism are based on the mistaken notion that most of the differences we observe in human intelligence and behavior are inherited and fixed. This completely ignores the fact that almost every aspect of our behavior is influenced to an extraordinary extent by our experiences and by what we learn from other people—from our culture. All of human history is a story of cultural change that as far as we can tell has required no genetic change whatever. This process may be called "cultural evolution" but it has essentially nothing to do with genetic, biological evolution. There is no evidence that we are more intelligent than the Greeks or Egyptians, or that any of the differences in our way of life are due to different genes. Similarly, there is not the slightest evidence that any of the cultural differences among peoples, which are the subject of anthropology, are due to genetic differences. They are due to the development of different cultural traditions that are passed from generation to generation by learning.[2] There is, moreover, not the slightest evidence that there are genetic differences in intelligence (whatever that is) among races; on the contrary, we know that blacks and whites, for example, achieve the same average IQ scores if they are raised in similar cultural and educational environments.[3]

There is no justification at all for supposing that less technologically sophisticated peoples are intellectually inferior; not the slightest evidence that poor people, unintelligent people, or criminals are the way they are because of their genes; and no foundation whatsoever for thinking that society would be improved by institutionalized "natural selection" that would prevent them from reproducing. This is the verdict of modern genetics; we must also, of course, bring our ethical principles to bear on the question. Eugenic breeding, whether in Plato's vision of the ideal republic or Hitler's evil vision of the totalitarian state, is an intolerable violation of our standards of human rights.

The scientific objections to Social Darwinism are trivial, though, compared to its philosophical faults. Spencer, Rockefeller, Marx, and Kropotkin all fell into the same trap—the fallacy of naturalism. This is the belief that what is natural is good, and must therefore provide moral guidance for human conduct. It is the same fallacy that Rousseau fell into, the belief that this is the best of all possible worlds and therefore serves as the best possible standard for human behavior. It is the belief that what is, ought to be.

There is no philosophical foundation for such a belief,[4] yet it has been held by innumerable people, scientist and nonscientist alike. One of the major figures of evolutionary biology in this century, Julian Huxley, was addicted to it. Huxley believed that evolution was a history of progress toward higher levels of order and consciousness, culminating in man. Man "is the agent through whom evolution may unfold its further possibilities." Thus "the ultimate guarantees for the correctness of our labels of rightness and wrongness are to be sought for among the facts of evolutionary direction." It is possible, then, says Huxley, to develop a philosophy of evolutionary ethics based on a few main principles: "that it is right to realize ever new possibilities in evolution, notably those which are valued for their own sake; that it is right to respect human individuality and to encourage its fullest development [because variation is the prerequisite for evolutionary change]; that it is right to construct a mechanism for further social evolution which shall satisfy these prior conditions as fully, efficiently, and as rapidly as possible."[5]

I cannot find fault in Huxley's liberal ideals; but his justification for these on the grounds of evolution strikes me as the same kind of philosophical hogwash as Rockefeller's justification of economic monopoly. At no time has anyone ever explained why human behavior should be modeled on nonhuman nature. Science discovers facts about nature. Those facts simply are; whether they ought to be is entirely irrelevant to their existence. Should stars explode into supernovas? Are earthquakes ethical? Should dinosaurs have become extinct? Should bedbugs evolve? These are nonsensical questions. Either earthquakes occur and bedbugs evolve or they do not. There is no moral imperative in the revolution of the planets telling each of us to stay in our appointed station in life. Nor is there any moral imperative in evolution telling us that we ought to progress, nor any deity called Natural Selection telling us that we ought to compete. Natural selection may be a "law of nature"; but a scientific law, like the "law of gravity," is merely a description of a regularity in natural processes, not a rule of conduct.

Many scientists, including many evolutionists, have seen that science is no guide to ethics. Perhaps the most eloquent statement has been *Evolution and Ethics*, an essay by Thomas Henry Huxley, Julian's grandfather.[6] Known as "Darwin's bulldog," the elder Huxley was Darwin's chief spokesman, and a far clearer thinker than his grandson. "Cosmic evolution may teach us how the good and the evil tendencies of man may have come about," he wrote; "but, in itself, it is incompetent to furnish any better reason why that which we call good is preferable to what we call evil than what we had before. Some day, I doubt not, we shall arrive at an understanding of the evolution of the aesthetic faculty; but all the understanding in the world will neither increase nor diminish the force of the intuition that this is beautiful and that is ugly." Huxley goes on, in this and an associated essay, to remark that a garden, or any other creation of human art and civilization, is maintained only by holding at bay the natural forces that tend to destroy it. The "survival of the fittest" is part of the natural forces that constitute the "cosmic process," but it is directly opposed to everything that we call civilization: "The influence of the cosmic process on the evolution of society is the greater the more rudimentary its civilization. Social progress means a checking of the cosmic process at every step and the substitution for it of another, which may be called the ethical process; the end of which is not the survival of those who may happen to be the fittest, in respect of the whole of the conditions which obtain, but of those who are ethically the best." Finally, "Let us understand, once for all, that the ethical progress of society depends, not on imitating the cosmic process, still less in running away from it, but in combating it."

Where, then, shall we find our sources of ethics and morals, our standards of right and wrong? Not in nature. If there is one moral lesson to be learned from evolution, it is that all of nature is utterly mechanical and amoral, and that the concepts of right and wrong, good and evil, are entirely limited to the human sphere. The standards of morality must be found either in the commands of a moral deity or in ourselves. Neither of these, of course, is very satisfactory. Every religion has different ethics, delivered to us not by its deity directly, but by human interpreters whose interpretations are open to argument—to say the least. And the evils committed in the name of God or the gods make a sad history.

The alternative is to devise our own ethics. This is the philosophy of humanism, the true target of creationists and other fundamentalist Christians. To them, evolution is evil because it seems to remove God from the universe, leaving humanism as the only source of morality. And since they don't like a lot of modern morality, they see evolution as the source of all modern evils. They forget, or do not know, that every evil we see today was part of Western society long before anyone conceived of evolution; they seem not to understand that science says nothing whatever about whether God exists or not; they cannot comprehend that the leaders of most of the major religions in America can accept the idea of evolution without any qualms. They do not recognize that humanism as a philosophy has many foundations, and does not need evolution to support it. Evolutionary biology cannot help us choose one philosophy of ethics over another. Conversely, ethics and morality have no place in the science classroom; nor does religion in any guise.

A nd this brings us to the flash point of the whole controversy: In the teaching of science, should creationism be given equal time with evolution? Do the creationists have a point in saying that both creation and evolution are rival scientific theories, neither of which has been proven? If so, is it not best, in the interest of free intellectual inquiry and academic freedom, to present both sides equally and let the student choose between them?

Let us understand, first, that no scientist says that children should be ignorant of the story of Genesis. It is, if nothing else, a beautiful allegory and an integral part of the cultural history of Western civilization. To be ignorant of it is to be severed from our heritage. There is, however, no chance that a child will be deprived of it. The opportunities for learning the Judaic and/or Christian tradition are abundant. But that is peripheral to the point at issue. Scientists object to teaching creation as a scientific theory, in science classes. By all means, teach Genesis in Sunday school, or in classes in history, comparative religion, or Western civilization. But creation has no place in a science class because it is not science.

Why not? Because creationism cannot offer a scientific hypothesis that is capable of being shown wrong. Creationism cannot describe a single possible experiment that could elucidate the mechanics of creation. Creationism cannot point to a single piece of scientific research that has provided evidence for any supernatural intervention into natural law. Creationism cannot point to a single prediction that has turned out to be right, and supports the creationist case. Creationism cannot offer a single instance of research that has followed the normal course of scientific inquiry, namely, independent testing and verification by sceptical researchers. Creationism cannot serve as a stimulus to scientific research, because it has no research program, no hypotheses, no predictions. Creationists can point to no source of their theory, no basis for their claims, other than the authority of the Bible. Science consists of posing testable, falsifiable hypotheses; making predictions about what is not yet known; performing critical experiments or observations that can disprove certain alternative hypotheses and lend credence to others; seeking explanations in natural rather than supernatural causes; trying to falsify hypotheses rather than to prove them; remaining sceptical until independent investigators are able to corroborate new claims; and subjecting one's ideas and data to the merciless criticism of other scientists. Creationism has none of these qualifications.

Suppose creationism had equal time in science classes. What would be taught? If creationists teach that the universe and all its inhabitants were suddenly created a few thousand years ago, and that all of extinction and all of geology were caused by a universal flood, what more can they say? Shall they provide scientific evidence that explains why blue-green algae are in the lowest geological strata and flowering plants in the uppermost? Shall they explain, in terms of modern biology, how a million or more species of animals fit into the ark? Shall they provide evidence from modern physics that explains away the fact that we can perceive light from stars that are billions of light years away, and took billions of years to get here? Shall they provide a testable hypothesis to explain the genetic similarity of apes and humans? Will they describe experiments that elucidate the mechanisms of creation, as geneticists have for the mechanisms of evolution? You will seek in vain for answers in *Scientific Creationism* or any other creationist books, and find instead only attacks on evolution, contorted explanations of how some phenomena could have been produced according to God's plan, and the occasional admission that creationist theories cannot be tested scientifically. "We cannot discover by scientific investigations," as creationist Duane Gish says, "anything about the creative processes used by the Creator."[7]

The business of science teachers is to teach science, not every conceivable theory that can be cooked up. Inheritance of acquired characteristics still stands as a theory of inheritance (until recently part of "scientific" dogma in the Soviet Union), but it is a discredited theory, and science teachers are under no obligation to include it in their courses. Neither are they obliged to teach the Ptolemaic theory of the solar system, or the theory that the earth is flat, even though there is a Flat Earth Society that holds this belief. Science teachers are obliged to teach science as it is currently understood by the community of practicing scientists in the field—not outmoded ideas that have been rejected by the world of science.

What an extraordinary perversion of science and education it would be if the teaching of creationism were demanded by law! The only example I know of in which a state has dictated what will be taught in the name of science is in the Soviet Union, where Lysenkoist genetics was state dogma. Nowhere in the United States does any law dictate what will be taught in science classes. Geologists do not need a law to help them convince people of the reality of continental drift; psychologists don't require the state to support their theories of human behavior; no science requires the force of law to fight its battles for it. If creationism were really science, it would stand or fall on the merits of its theories and research. It wouldn't need legislation to ram it down people's throats.

If evolution is a "theory," though, should it not be taught as a theory, rather than as dogmatic truth? By all means, but then students should know what a scientific theory is. They should learn that it is a coherent body of explanatory principles, not a mere speculative hypothesis. They should learn that chemists work with atomic theory and physicists with the Newtonian theory of forces. They should realize that no science consists of immutable truths, that every science consists of provisionally accepted hypotheses. They should realize that physics, chemistry, geology, and every other science have a history of argument and changes in theory, just as evolutionary biology has, and that the progress in each of these sciences consists of challenging established views with new ideas and experiments.

There is no justification for teaching creationism in the science classroom. But if it were taught, would it be subjected to the same critical analysis as the creationists insist should be brought to bear on evolution? Would the creationists permit a book such as this one to be used as one of several alternative texts, in pursuit of fair and open inquiry? Would they be willing to accept a science text that instead of merely presenting the evidence for evolution went so far as to expose the inconsistencies and absurdities of the creationist theory? And would they agree that all sciences should be taught on the same grounds of fairness? For if a scientifically untestable and discarded theory such as creation is taught in biology, I claim it is as logical to insist that astrology be taught as an alternate theory of psychology, extrasensory perception as a theory of physiology, alchemy as a theory of chemistry, and, indeed, divine providence as a theory of physics and astronomy. If evolution is to be taught as a mere theory that must compete with alternative explanations that have no scientific foundation, then the same principles should apply to all the sciences.

The fundamentalists who are opposed to evolution must be opposed, in the final analysis, to all of science. For their fight is a confused battle against materialism, and all of science is by its nature materialistic. Physicists do not explain the movement of heavenly bodies or the decay of atoms into elementary particles by recourse to the divine will. They rely instead on gravitation, quanta, and electromagnetic waves—material explanations for material events. Quite aside from Darwin's influence, biology has purged itself of the idea that life consists of a vital force or *élan vital;* all life processes are pure chemistry, and it is in terms of chemistry that the significant advances in biology are now being made. Nowhere in science is there any room for spiritual or nonmaterial explanation; the entire world view of the devout fundamentalist, based on concepts of purpose, providence, and divine guidance, is excluded from the world of science.

But materialism as a foundation for scientific explanation does not require that we adopt materialism as a philosophy of life. Science does not speak to our emotions, nor offer us moral codes. I may apply the canons of materialistic science to all of physics, chemistry, and biology, but I can still be moved to tears by the heartrending beauty of Wotan's farewell in *Die Walküre*, to a love of humanity by Sarastro's great aria in *Die Zauberflöte*, to compassion by the spectacle of Lear and Cordelia, to justice and morality by the teachings of Christ and the Buddha. The materialism of our society—the cold calculation of material gain, the cynical indifference to suffering humanity in the name of patriotism, the despoliation of the environment in the name of progress—is a social philosophy born of the politics and economics of self-interest, not of science. Materialism as a social attitude and materialism as a method of analysis in science are two different things. Instead of confusing them, as the fundamentalists do, it is right and proper to oppose the one and affirm the other.

Fundamentalists object to teaching evolution on the grounds that it contradicts the student's innate consciousness of reality and so creates mental and emotional conflict; that it leads to an amoral philosophy of "might makes right"; and that it discriminates against people who believe in creation. I have argued that none of these claims has any real foundation. But even if they had, would that justify teaching nonscience as science? Education is not designed to reinforce whatever innate consciousness of reality a student may have. Many of our supposedly innate beliefs—which in fact are generally merely learned from parents and peers—are wrong, and it is the job of education to correct them. Learning that the earth revolves around the sun, and that the universe is billions of light years in diameter, can be emotionally most disquieting, but it is important to face these facts, however much they may contradict our sense of reality. Learning that evolution has occurred has not deprived anyone I know of a sense of purpose, but the meaning of life is in any case a subject for personal reflection, not science. Evolution can lead to amorality only if we are taught to slavishly follow nature as a model for our behavior; but science should teach no such lesson.

What, on the other hand, are the implications of teaching creation as an alternate scientific theory? First, any creationist teaching in a public school necessarily violates the constitutional separation of church and state. However the creationists may seek to expunge any direct reference to Genesis from their books, they are forced to rely on a personal God to explain the origin of the universe and of living things. A creationist theory of the origin of species must have more content than the mere statement "living things were created in their present form by supernatural agencies." It must, if it is to satisfy any student, have an explanation that accounts for the world and its inhabitants, and must therefore go on to characterize those unnamed "supernatural agencies" as omnipotent, omniscient, and so forth—the characteristics attributed to the Judeo-Christian God. The creationist theory must go even further and explain all the geological observations that embarrass it—and so come up with a catastrophic flood story. The Genesis story, then, or something very much like it, is necessarily inherent in "scientific creationism," as of course it must be, because the entire motivation of the creationist movement is the defense of their literal reading of Genesis. The creationists do not accept any of the creation stories of other world religions, and would not tolerate for a moment the idea of presenting Chinese, Zulu, or Navajo creation stories on the same footing as the Biblical version. And of course the textbooks on scientific creationism that the fundamentalists want used in science classes are full of references, page after page, to an omniscient, omnipotent, personal, moral Creator who providentially preserved a remnant of humanity after a worldwide flood, near Mount Ararat. This is transparently the Genesis story, and no amount of equivocation can disguise the fact that the fundamentalists want this story, and this story only, presented as scientific truth.

Because creationism violates the separation of church and state, bills that mandate equal time for creation are strongly opposed by religious, political, and educational leaders who have no vested interest in evolution but perceive the creationist movement as a threat to our constitutional principles. Scientists oppose creationism on these grounds too, and on others. Grant creationism any status as a scientific theory and the very concept of science becomes meaningless. If creationism is science, then science does not consist of careful observation and experiment, of remorseless challenges to accepted ideas, of the exploration of natural causes, of the uniformitarian principles that guide all scientific investigation and thought. The child who learns that living things are created and directed by unknown and unknowable forces that can alter natural laws at will cannot apply the rules of scientific logic and procedure in any other field, from physics to psychology. He or she can only learn that the human quest for knowledge is doomed to failure by the willful whim of an inscrutable agent who acknowledges no natural law. Creationist theories rest not on evidence that can withstand the sceptical mind, but on wishful thinking and the Bible, the voice of authority which is the only source of creationist belief. To treat creationism as if it were respectable science is then to encourage credulity, gullibility, submission to authoritarian dogma, and the primacy of desire over evidence in shaping our view of reality.

Scientific creationism is an intolerable assault on education not merely because it is the antithesis of reason, but because it is opposed to the very foundation of true education: intellectual honesty. Surely education should teach the courage to weigh evidence and draw conclusions dispassionately, and to recognize their consequences, however hard or distasteful. Scientific creationism teaches, instead, the standards of the Madison Avenue marketplace: how to further your aims by guile, seductive catch phrases, selective quotation of evidence. Like the purveyors of cigarettes, laetrile, nuclear superiority, and instant spiritual enlightenment, scientific creationism teaches by its tactics more than by its words: truth is not the object of brave and honest search. Truth is whatever you can convince people it is. But to accede to these standards in education is to teach dishonesty and cowardice.

These, finally, are the social implications of the creation/evolution battle. Learning about evolution is not so important in itself as it is a context for learning how to think: how to derive conclusions logically, how to evaluate evidence, how to settle for tentative answers and replace them as better answers come along, how to see past superficial appearances to the reality beneath, how to question tradition and authority. It is no accident that the creationist movement is an arm of a larger political movement, the New Right, that strives to replace the pluralism and open debate in our society with its version of absolute, unquestioned truth. The New Right feeds on absolute answers and absolute adherence to its beliefs. It finds justification for its social and political positions in one place: its literal and authoritarian interpretation of the Bible. It can sway people to its side only by inculcating in them the same kind of absolutism and submission to authority.

The threat of evolution lies in its challenge to the uncompromising world view of the ultraconservative right. Question the literal truth of the Bible and you question the only authority on which the New Right can draw to support its uncompromising positions. The battle over evolution is a battle over the future of science in this country; but more deeply, it is a political contest between those who would constrain minds and those who would liberate them. The history of fascism shows how important it is to decide between an education that trains people to accept the platitudes of authority and the appeal to emotion, and one that develops the habit of informed scepticism and rational decision. The social role of science is, in part, to provide for technological progress; but its more important role is to train us in the habit of rational thought, and to encourage scepticism and free inquiry. Looked on in its larger context, the assault on evolution is an assault on science, and on political freedom.

For most of its history, the human race has dwelt in the shadows of ignorance and superstition. In their fear of the unknown, peoples have invented myths to explain the vicissitudes of fortune. Only in the last century or so has humanity been able to find the causes of disease, suffering, the movements of the earth and the tides and the stars; only in the last century have we been able to use physics, chemistry, and biology to our benefit. Until rationalism and science arose and made this knowledge possible, humanity was as a child full of wonder, dread, and innocence. But if maturity is the loss of innocence, it is also the attainment of knowledge, independence, and responsibility. We cannot go back to that mythical age of innocence for which the fundamentalists yearn, and we would not wish to return to the childlike fantasies and ignorance in which they still labor. We cannot afford to. Science and the use of reason have loosed upon us forces that have served us well, and others that will destroy us unless they are tamed by the same use of reason that gave them birth.

But what reason has achieved, it has achieved in the face of the opposition of authority and tradition. If authority had had its way, we still would not know what Galileo saw through his telescope. It is inherent in rationalism, and in the apotheosis of rationalism that is science, that tradition and authority must give way to new worlds of thought. The voice of authority must be forever threatened by the rational mind that dares to doubt, and so rationalism is our best defense against political tyranny. Authority draws its strength from its reliance on law and force; rationalism from its faith in the human mind.

APPENDIX

SOME CREATIONIST ARGUMENTS, AND SOME APPROPRIATE RESPONSES

I list here some of the common creationist claims, each with a capsule counterargument. Most of these are developed fully in the body of the text.

I. Philosophical and Scientific Issues

1. *Evolution is outside the realm of science because it cannot be observed.*

    Most of science depends not on direct observation, but on testing predictions that derive logically from hypotheses. We do not know the structure of an atom or a DNA molecule from direct observation.

2. *Evolution cannot be proven.*

    Nothing in science is ever proven; we merely achieve greater and greater confidence in the validity of our hypotheses as more data support or fail to refute them.

3. *Evolution is not a testable hypothesis because it could not be refuted by any possible observation.*

    Many conceivable observations, such as mammalian fossils in Precambrian rocks, could refute the hypothesis of evolution.

4. *Evolution is a religion because it is based on unobservable processes and because it includes hidden concepts of ethics, values, and ultimate meanings.*

    Belief in evolution does not require faith, because the processes can in fact be observed or tested. If religion is defined as including any set of beliefs that touches on ethics and values, then evolution is not a religion, because it describes what has occurred, not what should occur.

5. *The order of the universe, and the adaptations of organisms to their environments, are evidence of intelligent design and purpose.*

Order can be observed to arise from the action of natural laws and physical processes, and is not evidence of design.

II. Natural Law

6. *Because the second law of thermodynamics holds that entropy (disorder) increases, all change must be degenerative, and greater complexity could never have evolved.*

The second law applies only to closed systems. Organisms, which exist in open systems, can capture energy and use it to build greater chemical order, and they do so all the time.

III. Biological Evolution

7. *It is infinitely improbable that even the simplest life could arise from nonliving matter.*

The formation of any particular nucleic acid sequence by chance is very improbable, but the chance of forming one or another viable form is very high.

Under conditions resembling those on the prebiotic earth, simple organic molecules actually form from elementary constituents (ammonia, methane, etc.), and assemble themselves into self-replicating nucleic acids which mutate and are altered in frequency by natural selection, all in the laboratory.

8. *Mutations are harmful, and do not give rise to new characteristics.*

Mutations that have "large" effects are usually harmful, but most mutations have small effects, and many of these are demonstrably beneficial under certain environmental conditions.

9. *Natural selection cannot create new characteristics; it is a conservative process that merely eliminates unfit mutants.*

First, most characteristics that appear in evolution aren't really new; most are changes in size, shape, developmental timing, or organization of preexisting characteristics. Second, natural selection acts as an editor, not an author: it shapes adaptive characteristics out of the chaotic new variations that arise by mutation and genetic recombination. Third, the evolution of important new characteristics by natural selection has been observed frequently, as in the development of new metabolic capacities in bacteria.

10. *Chance could not be responsible for the origin of complex organisms, which therefore bear evidence of design.*

    Although mutations arise by chance, they succeed or fail to become established in a species by natural selection, which is the antithesis of chance.

11. *Natural selection is an untestable, tautologous concept. The "fittest" are those who survive, who in turn are labeled the fittest.*

    Given an understanding of the relation of each of two forms to an environment, one can predict which will more readily survive and reproduce; thus testable predictions about natural selection can be and have been made. It is also possible to find genetic changes that are not caused by natural selection, which shows that natural selection is not automatically invoked, by circular argument, to explain all evolutionary change.

12. *A new structure would not have any selective advantage when it first appears in a rudimentary condition, and so could not develop at first by natural selection.*

    Even complex organs such as the eye are often represented by less complicated structures in more "primitive" species, in which they are fully functional. Moreover, not all changes are brought about by natural selection; some features become elaborated because of their correlation with the growth of other features, and only then become useful.

13. *No fossils have been found that exemplify an incipient structure evolving into a subsequently useful feature.*

    This is not true; the fossil record provides many examples. One case, described in the text, is the incipient ridges on the teeth of early horses that subsequently became greatly elaborated as an adaptation for grinding vegetation.

14. *If gradual evolution had occurred, there would be no gaps among species, and classification would be impossible.*

    Many disparate organisms are connected by intermediate species, and in such cases classification *is* arbitrary. In many other cases gaps exist because of extinction.

15. *Despite a rich fossil record, transitional intermediates between ancestors and descendants are not found in the fossil record.*

    The modern genetic theory of evolution holds that adaptation to new conditions proceeds rapidly, so that few intermediates are likely to be found. Even so, many cases of evolutionary transitions from one species to a related species are known in the fossil record, and the gradual origin of several groups (e.g., ammonites from bactritid cephalopods, mammals from therapsid reptiles, horses from condylarths) is well documented by the fossil record. *Archaeopteryx*, contrary to creationists' claims, is not a full-fledged bird, but a reptile with a few avian characteristics such as feathers.

16. *If evolution is true, why should "living fossils" such as the coelacanth and the horseshoe crab have evolved little for hundreds of millions of years?*

    If a species is adequately adapted to its environment, there is no reason to expect it to continue to evolve new adaptations.

17. *Homologous anatomical structures, and similarities in embryonic development, are examples of a common design used by the Creator, not of common ancestry.*

    Of course anyting can be "explained" by the Creator's desires, since we have no way to obtain information about the Creator. But many homologous structures make no adaptive sense, and do not conform to any optimal design that we can understand. There are no design constraints that require sharks and humans to have similar embryos and yet develop into completely different organisms.

18. *Vestigial structures are not vestigial but functional.*

    There is not the slightest reason to think that many vestigial structures, which violate rational design, have any function. The pelvic bones of pythons and the rudimentary wings of many insects have no known function, and related species of snakes and insects lack them altogether.

19. *The fossil record is not an objective time sequence, because it is already assumed that evolution occurred, and only then are the rocks "ordered" by their fossil contents.*

    In fact, the geological ordering of fossil strata was made by preevolutionary geologists who believed in creation. Moreover, radioactive dating and other methods are also used to establish a relative geological sequence.

20. *There is no proof that radioactive decay occurs at a constant rate, and hence no proof that the earth is billions of years old.*

    The theory of physics, and the failure to find any factors that could alter rates of radioactive decay, provide a solid foundation for radioactive dating. Radioactive dates are consistent with many other sources of evidence indicating that the age of the earth, solar system, and universe must be measured in billions of years.

21. *Geologists are abandoning the principle of uniformitarianism; as a result, explanation of geological features, such as the fossil record, by catastrophes such as a universal flood, is equally plausible.*

Uniformitarianism holds that only presently observable natural forces have operated in the past, as they do today, although their rates can vary. Even at the most rapid rates known to exist, natural forces cannot account for geological features such as continental drift or sedimentation except on a time scale of many millions of years. The ordering of fossils, and many other features of the fossil record, cannot conceivably be explained either by a single catastrophe or a series of catastrophes.

22. *No one can know events that occurred before there were people to observe and record them; thus past evolutionary events can never be known.*

Direct observation is not the only source of reliable evidence; and in fact direct observation often provides untrustworthy evidence. Past events can be reliably inferred by logical deduction, using a variety of methods that are common to all historical sciences.

IV. Human Evolution

23. *There are no intermediate fossils between humans and apes. Australopithecus walked on all fours like modern apes, and had an apelike skull. It was merely an ape.*

The anatomy of australopithecines, including the recently discovered form called *Australopothecus afarensis*, clearly indicates an erect posture, coupled with a rather apelike skull with a small brain. Later hominid fossils clearly approach modern humans in successive steps, in several features such as brain size and dentition. Associated stone tools also show progressive complexity of design.

24. *If races diverged in skin color and other trivial features, why haven't they diverged in intelligence, which is so critical to survival?*

Different characteristics experience different patterns of selection. Intelligence was, very likely, evolved to its modern degree in ancestral human populations before they spread into new areas, where divergence in skin color may have been locally adaptive (or may have occurred by genetic drift). Because of the high survival value of intelligence, it would experience stabilizing selection to maintain the same high level everywhere. This is not at all the "unsolved puzzle" the creationists make it out to be.

25. *Evolutionists in Darwin's day were racists, and the very concept that*

each race has taken a long time to evolve leads to racism. Racism is the concept that each race has a long, separate evolutionary history.

Racism is, on the contrary, the social attitude that holds that the characteristics (especially personal and social) of an individual must conform, a priori, to those that are thought (usually without evidence) to be typical of the race to which the individual belongs, without taking into account the existence of individual variability. It is not founded at all on the concept of evolution. Evolutionary divergence in characteristics such as skin color does not at all imply that other characteristics such as intelligence have diverged. If nineteenth-century evolutionists were racist (and not all were), they merely shared the nonscientific assumptions of their society.

26. *There is an unbridgeable gap in intelligence and emotions between humans and all other species that cannot be accounted for by evolution.*

The mental characteristics of *Homo sapiens* are indeed developed to a far greater degree than seems to be true of any other species, yet most of the mental faculties we have seem to be present in more rudimentary form in other primates and mammals. If cognition, emotion, etc. have a physical basis in the brain, which is the working assumption of psychology, then the physical basis for cognition and consciousness could evolve, just as other physical features do.

V. General Issues

27. *Quotations from many eminent evolutionists show that biologists have abandoned the concept of gradual evolution by natural selection. Darwin was wrong, and the entire study of evolution is in disarray.*

Most of the quotations cited by creationists in justification of this posture are from evolutionists who claim that (a) gradual transitions between species are uncommon in the fossil record; (b) many characteristics of species do not appear to be adaptations; (c) evolution may proceed by large mutational changes as well as small ones; (d) the theory of natural selection doesn't explain certain major events and trends in the overall history of life.

Point (a) is taken by evolutionists to mean that evolution often occurs very rapidly, on a very local geographic scale, as is often demonstrably the case. This in no way contradicts traditional evolutionary theory. Point (b) has also been recognized by evolutionists ever since Darwin; many factors besides natural selection govern the direction and rate of evolutionary change. Point (c) is largely a matter of defining some mutations as "large" in effect and others as "small"; there is, in reality, a complete spectrum of effects. Point (d) is an important recognition that the forces of mutation and natural selection that adapt a species in the short run may not be correlated with its chance of survival in the long run. The factors that cause ultimate extinction of some species and not others require more research than they have received, and a higher-level theory that includes traditional neo-Darwinian theory may be required. The evolutionists quoted do not deny the validity of evolutionary theory but are seeking to expand it to cover a broader range of phenomena. In most respects, Darwin is perceived to have been right, but a lot of the details of how evolution operates are still being worked out. This is evidence not of disarray in the science, but of healthy progress in the search for more complete explanations.

28. *Evolution should not be taught because it leads to a materialistic, amoral philosophy of "might makes right."*

The philosophical or ethical implications of any scientific statement do not bear on its scientific validity; neither they nor our desires make the statement either right or wrong. Ethics and philosophy are not part of science, and no moral lessons about how we should behave can logically be deduced from evolution or any other science. Scientific statements, whether in physics or biology, are materialistic in the sense that they explain natural phenomena by natural, material causes; they are amoral in that they describe what is, without making value judgments about whether it ought or ought not to be. The answers to ethical and moral questions must be found outside of science.

RECOMMENDED FURTHER READINGS

Although I am averse to supporting the creationist cause financially by buying their publications, I must recommend some creationist literature as the surest antidote to believing the creationist line. Many publications of Creation-Life Publishers, P.O. Box 15666, San Diego, California 92115 will do nicely. I have quoted most extensively from *Scientific Creationism* (Public School Edition), edited by H. M. Morris, and from *Evolution: The Fossils Say No!* by D. T. Gish. *The Troubled Waters of Evolution,* by H. M. Morris, contains almost the same material but is more straightforwardly religious in tone. *The Genesis Flood,* by H. M. Morris and J. C. Whitcomb, presents a "scientific exposition" of creation and the flood. *The Early Earth,* by J. C. Whitcomb, "shows how God's Word refutes any type of theistic evolution."

Creation/Evolution (P.O. Box 5, Amherst Branch, Buffalo, New York 14226) is a quarterly publication dedicated to promoting evolutionary science and covers current creationist political activities. Issue 7 (Winter 1982) has a useful article, "Answers to the Standard Creationist Arguments," by K. Miller.

The American Biology Teacher (11250 Roger Bacon Drive, Reston, Virginia 22090) is a must for teachers and has had numerous useful articles, including:

Alexander, R. D. 1978. "Evolution, Creation, and Biology Teaching." ABT 40(2):91–104.

Callaghan, C. A. 1980. "Evolution and Creationists' Arguments." ABT 42(7):422–25.

Miller, K. R. 1982. "Special Creation and the Fossil Record: The Central Fallacy." ABT 44(2):85–89. (An important exposition of how the flood scenario is essential to the creationist argument, and why it is absurd.)

Hughes, S. W. 1982. "The Fact and the Theory of Evolution." ABT 44(1):25–32. (Includes a good capsule review of the coverage of evolution by high school textbooks.)

On the legal front, the important decision in *McLean* v. *Arkansas Board of Education* was reprinted in full in *The American Biology Teacher* 44(3):172–79 (March 1982), and in *Science* 215:934–43 (February 19, 1982).

SOME OTHER USEFUL ARTICLES INCLUDE:

Brush, S. G. 1981. "Creationism/Evolution: The Case Against 'Equal Time.'" *The Science Teacher*, vol. 48, no. 4.

Asimov, Isaac. 1981. "The 'Threat' of Creationism." *New York Times Magazine*, June 14, 1981. Ideal for classroom discussion and for enjoyment of his spirited style.

Cloud, P. 1977. "Scientific Creationism—A New Inquisition Brewing." *The Humanist* 37:1.

Nelkin, D. 1976. "The Science-Textbook Controversies." *Scientific American* 234(4):33–38. Analyzes reasons for creationist and other attacks on science curricula.

Skow, J., et al. 1981. "The Creationists." *Science 81*, pp. 53–60 (December). A succinct summary of who they are and how they operate.

Brush, S. G. 1982. "Finding the Age of the Earth by Physics or by Faith?" *Journal of Geological Education* 30:34–58. A thorough analysis of radioactive dating and how the creationists deal with it.

BOOKS:

Stebbins, G. L. *Processes of Organic Evolution* (Englewood Cliffs, N.J.: Prentice-Hall, 1971). A short, elementary introduction to the theory of evolutionary change. For introductory college courses.

Stansfield, W. D. *The Science of Evolution* (New York: Macmillan, 1977). A somewhat more advanced textbook.

Dobzhansky, Th., F. J. Ayala, G. L. Stebbins, and J. W. Valentine, *Evolution* (San Francisco: Freeman, 1977).

Futuyma, D. J. *Evolutionary Biology* (Sunderland, Mass.: Sinauer, 1979). This and the preceding book are the most comprehensive current college textbooks on the subject.

Stebbins, G. L. *Darwin to DNA, Molecules to Humanity* (San Francisco: Freeman, 1982). This is one of the few recent books designed to explain evolution for the general reader, by one of the leaders in the field.

Cloud, P. *Cosmos, Earth, and Man* (New Haven: Yale University Press, 1978). A very readable nontechnical discussion of the history of the universe, the earth, and living things.

Johanson, D. and E. Maitley, *Lucy: The Beginnings of Mankind* (New York: Simon & Schuster, 1980). A lively account of one anthropologist's fossil discoveries and interpretation of human evolution; conveys a nice sense of what is and isn't known, and what to do in order to know more.

Gould, S. J. *The Panda's Thumb* (New York: Norton, 1980). Superb essays about evolution and science by perhaps the best writer in science today.

Godfrey, L., ed., *A Century After Darwin* (Boston: Allyn & Bacon, 1983). A moderately technical collection of essays treating the development of evolutionary thought since Darwin. Touches on creationism only slightly.

THE CREATIONIST DEBATE IS EXPLICITLY TREATED IN THE
FOLLOWING BOOKS:

Eldredge, N. *The Monkey Business* (New York: Washington Square Press, 1982). A short, very nontechnical treatment of the arguments.

Newell, N. *Creation and Evolution: Myth or Reality?* (New York: Columbia University Press, 1982). For the general reader, treating especially paleontological aspects.

Kitcher, P. *Abusing Science: The Case Against Creationism* (Cambridge,

Mass.: MIT Press, 1982). A philosopher's detailed analysis of creationist arguments.

Godfrey, L., ed., *Scientists Confront Creationism* (New York: W. W. Norton, 1983). A collection of nontechnical essays, by scientists from many fields, that refute creationist arguments on astronomy, geology, biology, and anthropology.

NOTES

CHAPTER ONE

1. Progress reports on the state of the creation-evolution controversy appear, among other places, in *Science*. See for example, *Science* 214:1101 (December 4, 1981), *Science* 214:1224 (December 11, 1981), *Science* 215:934 (February 19, 1982).
2. *San Francisco Sunday Examiner and Chronicle*, March 8, 1981.
3. This history is given in greater detail by D. Nelkin, *Science Textbook Controversies and the Politics of Equal Time* (Cambridge, Mass.: MIT Press, 1977).
4. Application form for the Creation Research Society, Ann Arbor, Michigan.
5. D. Nelkin, *op. cit.*
6. *Village Voice*, October 14–20, 1981.
7. *San Francisco Sunday Examiner and Chronicle*, March 8, 1981.
8. *Village Voice, ibid.*
9. R. E. Kofahl and K. L. Segraves, *The Creation Explanation* (Wheaton, Ill.: H. Shaw, 1975).
10. H. M. Morris, *Biblical Cosmology and Modern Science* (N.J.: Craig Press, 1970), p. 71.
11. H. M. Morris, *Remarkable Birth of Planet Earth* (Minneapolis, Minn.: Dimension Books, 1972), p. 66.
12. H. M. Morris and J. C. Whitcomb, *The Genesis Flood* (Nutley, N.J.: Presbyterian and Reformed Publishing Co., 1961).
13. H. M. Morris, ed., *Scientific Creationism* (San Diego: Creation-Life Publishers, 1974). Henceforth referred to as *Scientific Creationism*.

14. *Creation: Acts/Facts/Impacts* (San Diego: Creation-Life Publishers, 1974), p. 183.

15. D. T. Gish, *Evolution: The Fossils Say No!* (San Diego: Creation-Life Publishers, 1974). Henceforth referred to by title.

16. *McLean* v. *Arkansas Board of Education.* U.S. District Court Judge William R. Overton's detailed and far-reaching opinion, issued January 5, 1982, is reprinted in full in *Science* 215:934 (February 19, 1982).

17. As quoted in opinion by Judge William R. Overton in *McLean* v. *Arkansas Board of Education.*

18. H. M. Morris, *Introducing Scientific Creationism into the Public Schools* (San Diego: Institute for Creation Research, 1975). The paragraph reads in full: "The teacher should then be encouraged (not required) to use this information in his or her classes. [Quoted passage.] For example, when treating a subject such as human origins, the teacher can balance the usual evolutionary discussion of Ramapithecus, Australopithecus, Neanderthal, etc., by citing the creationists' evidence that such fossils are invariably either of apes or of men, with no true and unquestioned intermediates between men and apes. Such a discussion need not deal with such theological topics as the divine purpose for man, but only with the factual evidence concerning the unique physical and mental characteristics of men."

19. H. M. Morris, "The Anti-creationists," *Impact*, no. 97 (San Diego: Institute for Creation Research, 1981). The paragraph continues "Creationists do *not* want the Biblical record of creationism taught in the public schools, but only the general creation model as a viable scientific alternative to the general evolution model."

20. *Scientific Creationism*, p. 12. The paragraph reads: "The creation model . . . in the model." [As presently written.]

21. *Ibid.*, pp. 19–20. The paragraph begins, "In justification of his own decision, however, the creationist utilizes the scientific law of *cause-and-effect.*"

22. *Ibid.*, p. 33. Paragraph quoted in full.

23. *Ibid.*, p. 32. "The earth, with its unique hydrosphere, atmosphere, and lithosphere is, so far as all the actual evidence goes, the only body in the universe capable of sustaining higher forms of life such as man. This, of course, is exactly as would be predicted from the creation model. [Quoted material.]"

24. *Ibid.*, p. 35. The paragraph continues, "Even the evolutionist recognizes that man is the highest product of the cosmic process. 'In man is a three-pound brain which, as far as we know, is the most complex and orderly arrangement of matter in the universe.'" The sentence quoted is from Isaac Asimov, *Smithsonian Institute Journal* (June 1970). Note that this conclusion is not at all the same as the man-ward orientation of creation postulated in the beginning of the paragraph.

25. *Ibid.*, p. 137. Paragraph quoted in full.

26. *Ibid.*, pp. 111–12. Quoted material is preceded by the words "The creation

model, on the other hand, must interpret the column" and is otherwise quoted in full.

27. *Ibid.*, p. 117. Paragraph continues, "The uniformitarian will of course question how such a cataclysm could be caused, and this will be considered shortly, but for the moment simply take it as a model and visualize the expected results if it should happen today." The "expected results" include those described in the text of my description of this model. The promised causes of such a cataclysm include eruptions of the earth's crust that released vast quantities of water stored within the earth, and precipitation of a vast blanket of water vapor that enveloped the earth.

28. *Ibid.*, p. 119. Paragraph begins, "Similarly these higher animals (land vertebrates)"

29. *Ibid.*, p. 187–88. Paragraph begins, "The creation model explains these same data in a completely different context, of course, but the data fit the creation model at least as well as the evolution model." The data referred to are the existence of Stone Age cultures whose members have the same potential skills as humans in other cultures.

30. *Ibid.*, pp. 188–89. Quoted in full.

31. *Ibid.*, p. 201. Paragraph continues, "Such decisions are, of course, very important decisions, and each individual is responsible, both to himself and to his Creator (if indeed creationism is true), to face them. They have profound consequences, both throughout, and even beyond, one's life."

32. *Ibid.*, p. 62. Quoted in full.

33. *Ibid.*, p. 52. Paragraph begins, "In other words, the phenomenon of variation and natural selection, rather than explaining evolution in the way Darwin thought it did, is really a marvelous example of the creationist's principle of conservation in operation. That is, a fundamental prediction from the creation model is that, [quoted material]."

34. *Ibid.*, p. 51. Paragraph begins, "Normal variations were later found to be subject to the rigid Mendelian laws of inheritance, representing nothing really novel, but only characters already latent within the genetic system. Modern molecular biology, with its penetrating insight into the remarkable genetic code implanted in the DNA system, has further confirmed that [quoted material]."

35. *Evolution: The Fossils Say No!*, p. 32. Paragraph begins, "We must here attempt to define what we mean by a basic kind."

36. *Ibid.*, pp. 34–35. Full text of paragraph is "In the above discussion, we have defined a basic kind as including all of those variants which have been derived from a single stock. We have cited some examples of varieties which we believe should be included within a single basic kind. We cannot always be sure, however, what constitutes a separate kind. The division into kinds is easier the more the divergence observed. It is obvious, for example, that among invertebrates the protozoa, sponges, jellyfish, worms, snails, trilobites, lobsters, and bees are all different kinds. Among the vertebrates, the

fishes, amphibians, reptiles, birds, and mammals are obviously different basic kinds." Gish is perhaps not aware that "worms" are classified into eleven different phyla, most of which are as different from one another as from other major groups of animals; that the simplest sponges are little more than aggregations of cells that are almost identical to the choanoflagellate protozoans; that intermediate forms connect the jellyfish to the corals, which are as superficially different from jellyfish as lobsters are from trilobites.

37. *Scientific Creationism*, p. 14. The elided material reads "He knows, as part of his own experience of reality, that a house implies a builder and a watch a watchmaker. As he studies the still more intricately complex nature of, say, the human body, or the ecology of a forest."

38. *Ibid.*, p. 15. Quoted material, consisting of four numbered (5–8) sentences, is preceded by "Conversely, there are serious objections and harmful aspects to the present practice of teaching evolution exclusively as the only acceptable explanation of origins. Some of these problems are as follows: 1. It is discriminatory and unfair to those children and parents who, for whatever reason, believe in creation. 2. It is contrary to the principles of civil rights. 3. It is destructive of scientific objectivity, which requires fair examination of competing models as a basis for decision. 4. It is inimical to the principle of academic freedom for those teachers who desire to teach creationism but are inhibited from doing so by fear of academic reprisals." See note 2, Chapter 9.

39. H. M. Morris, in *Creation: Acts/Facts/Impacts*, p. 160. The paragraph begins, "In a day and age which practically worshipped at the shrine of scientific progress, as was true especially during the century from 1860 to 1960, such universal scientific racism was bound to have repercussions in the political and social realms."

40. M. E. Marty, *Prime Time* (August 1981).

41. D. T. Gish, in *Creation: Acts/Facts/Impacts*, p. 74. The full paragraph is "No doubt highly-competent, scientifically-trained, dedicated Christians could be found to augment our staff and undertake these projects. Our problem is the lack of funds necessary to provide for these additional staff members. We urge God's people to become sufficiently concerned about this cancer of evolution-oriented secular humanism that is destroying the minds and faith of our young people; then the necessary prayer and financial support will be provided to allow for the urgently-needed expansion of our staff."

42. *New York Times*, March 17, 1981.

43. *Ibid.*

44. *New York Times*, September 6, 1981.

45. "A Reporter at Large: A Disciplined, Charging Army," by Frances Fitzgerald; © 1981 in *The New Yorker Magazine* (May 18, 1981), p. 99.

46. R. Hofstadter, *Anti-Intellectualism in American Life* (New York: Vantage Books, 1962).
47. M. E. Marty, *op. cit.*
48. *Ibid.*
49. These quotations are from *A Compendium of Information on the Theory of Evolution and the Evolution-Creationism Controversy*, National Association of Biology Teachers, 11250 Roger Bacon Drive, Reston, Va. (1978).
50. D. Nelkin, *op. cit.*

CHAPTER TWO

1. Andrew Dickson White, *A History of the Warfare of Science with Theology in Christendom*, vol. 1 (London: Macmillan, 1896; reprint ed., New York: Dover, 1960).
2. A. O. Lovejoy, *The Great Chain of Being* (Cambridge, Mass.: Harvard University Press, 1936).
3. Much of this history is provided by J. C. Greene, *The Death of Adam: Evolution and its Impact on Western Thought* (Ames: Iowa State University Press, 1959).
4. A detailed history of this and other developments in evolutionary biology is given by Ernst Mayr, *The Growth of Biological Thought: Diversity, Evolution, Inheritance* (Cambridge, Mass.: Harvard University Press, 1982).
5. See D. L. Hull, *Darwin and His Critics* (Cambridge, Mass.: Harvard University Press, 1973).
6. *Ibid.*
7. E. Mayr and W. B. Provine, *The Evolutionary Synthesis* (Cambridge, Mass.: Harvard University Press, 1980).
8. Our modern understanding of the mechanisms of evolution is described in many books. Elementary textbooks include G. L. Stebbins, *Processes of Organic Evolution*, (Englewood Cliffs, N.J.: Prentice-Hall, 1971), and J. Maynard Smith, *The Theory of Evolution* (New York: Penguin Books, 1975). More advanced textbooks include Th. Dobzhansky, F. J. Ayala, G. L. Stebbins, and J. W. Valentine, *Evolution* (San Francisco: Freeman, 1977), and D. J. Futuyma, *Evolutionary Biology* (Sunderland, Mass.: Sinauer, 1979). Unreferenced facts and theories described in the text are familiar enough to most evolutionary biologists that they will be found in most or all of the references cited above.

CHAPTER THREE

1. S. J. Gould, *The Panda's Thumb* (New York: Norton, 1980).
2. A. Hampé, *J. Embryol. Exper. Morph.* 8:241 (1960).
3. E. J. Kollar and C. Fisher, *Science* 207:993 (1980).
4. W. M. Fitch and E. Margoliash, *Evolutionary Biology* 4:67 (1970).
5. P. J. Darlington, *Zoogeography: The Geographic Distribution of Animals* (New York: Wiley, 1957).
6. Continental drift and its consequences are described in J. T. Wilson, ed.,

Continents Adrift and Continents Aground: Readings from Scientific American (San Francisco: Freeman, 1976).

7. G. L. Stebbins, *Flowering Plants: Evolution Above the Species Level* (Cambridge, Mass.: Harvard University Press, 1974).

8. E. O. Wilson, F. M. Carpenter, and W. L. Brown, *Science* 157:1038 (1967).

9. M. Goodman, *Progress in Biophysics and Molecular Biology* 38:105 (1982).

10. E. Mayr and D. Amadon, Amer. Mus. Novitates no. 1496, American Museum of Natural History, New York, 1951.

11. See, for example, A. S. Romer, *Vertebrate Paleontology* (Chicago: University of Chicago Press, 1960). A recent discussion of intermediate series and gaps in the fossil record is provided by E. C. Olson, *Quart. Rev. Biol.* 56:405 (1981).

12. *Scientific Creationism*, p. 72.

13. D. L. Hull, *op. cit.*

14. S. J. Gould and R. C. Lewontin, *Proc. Roy. Soc. Lond.* 205:147 (1979).

15. R. B. Goldschmidt, *The Material Basis of Evolution* (New Haven: Yale University Press, 1940).

16. E. Mayr, in S. Tax, ed., *The Evolution of Life* (Chicago: University of Chicago Press, 1960), p. 349.

17. S. J. Gould, *Ontogeny and Phylogeny* (Cambridge, Mass.: Harvard University Press, 1977). See also G. Oster and P. Alberch, *Evolution* 36:444 (1982).

CHAPTER FOUR

1. J. C. Greene, *op. cit.*

2. D. M. Raup, *Science* 213:289 (1981).

3. Methods of dating geological material are described by D. L. Eicher, *Geologic Time* (Englewood Cliffs, N.J.: Prentice-Hall, 1976).

4. Eicher, *op. cit.*

5. S. van den Bergh, *Science* 213:825 (1981).

6. *Ibid.*

7. The early history of the earth and the origin of life are treated by P. E. Cloud, *Cosmos, Earth, and Man* (New Haven: Yale University Press, 1978). See also *Scientific American* (September 1978).

8. *Ibid.*

9. A. S. Romer, *op. cit.*

10. J. Ostrom, *Biol. J. Linn. Soc.* 8:91 (1976).

11. A. S. Romer, *op. cit.*

12. *Ibid.*

13. For a general introduction to paleontological methods, see D. M. Raup and S. M. Stanley, *Principles of Paleontology* (San Francisco: Freeman, 1971).

14. G. G. Simpson, *The Major Features of Evolution* (New York: Columbia University Press, 1953).

15. G. G. Simpson, *Horses* (New York: Oxford University Press, 1951).

16. B. Kurtén, *Cold Spring Harbor Symp. Quant. Biol.* 24:205 (1959).

17. T. S. Westoll, in G. L. Jepsen, G. G. Simpson, and E. Mayr, eds., *Genetics, Paleontology, and Evolution* (Princeton, N.J.: Princeton University Press, 1949).

18. N. Eldredge and S. J. Gould, in T. J. M. Schopf, ed., *Models in Paleontology* (San Francisco: Freeman, Cooper & Co., 1972); also S. M. Stanley, *Macroevolution: Pattern and Process* (San Francisco: Freeman, 1979), and S. M. Stanley, *The New Evolutionary Timetable* (New York: Basic Books, 1981).

19. The difference in viewpoint between advocates of punctuated equilibria and gradualism may be just a matter of scale; see G. L. Stebbins and F. J. Ayala, *Science* 213:967 (1981), and S. Wright, *Evolution* 36:427 (1982).

20. Rapid, extensive genetic change in laboratory experiments is described in any evolution textbook; see also Chapters 7 and 8.

21. P. G. Williamson, *Nature* 293:437 (1981).

22. H. K. Erben, *Biol. Rev.* 14:641 (1966); also E. C. Olson, *op. cit.*

23. P. D. Gingerich, *Am. J. Sci.* 276:1 (1976).

24. D. E. Kellogg, *Paleobiology* 1:359 (1975).

25. A. S. Romer, *op. cit.;* E. C. Olson, *op. cit.*

26. The evolution of horses is described in great detail by G. G. Simpson in *Horses* (cited above). It has been claimed by a critic of evolutionary theory ("Nova" television program, November 1981) that the sequence of horse fossils was arranged by early workers to fit their preconceptions and does not actually fit the sequence in fossil deposits. Paleontologists Leonard Radinsky and James Hopson, authorities on the fossil record of the mammals, have told me that this claim is absolutely without foundation, and that recent research confirms Simpson's account in every essential detail.

27. D. T. Gish, *Impact*, no. 87 (San Diego: Institute for Creation Research, 1980). See also note 26.

28. G. G. Simpson, *The Major Features of Evolution* (New York: Columbia University Press, 1953), pp. 260–65.

29. *Ibid.*, p. 345.

30. See, for example, M. Calvin, *Chemical Evolution* (New York: Oxford University Press, 1969); R. E. Dickerson, in *Scientific American* (September 1978).

31. M. Eigen et al., in *Scientific American* 244:88 (April 1981).

CHAPTER FIVE

1. J. C. Greene, *op. cit.*, discusses early thoughts on human origins.

2. Anatomical and paleontological aspects of human evolution are treated in many works, e.g., D. Pilbeam, *The Ascent of Man* (New York: Macmillan, 1972).

3. G. G. Gallup, in R. H. Tuttle, ed., *IX Internat. Congr. Anthropol. Ethol. Sci., Primatology Session* (The Hague: Mouton Press, 1974).

4. This topic is summarized by S. J. Gould, *Ontogeny and Phylogeny* (Cambridge, Mass.: Harvard University Press, 1977).

5. M.-C. King and A. C. Wilson, *Science* 188:107 (1975).
6. V. M. Sarich and A. C. Wilson, *Science* 158:1200 (1967).
7. J. J. Yunis, J. R. Sawyer, and K. Dunham, *Science* 208:1145 (1980).
8. See D. Pilbeam, *op. cit.*, for most of this paleontological history.
9. D. C. Johanson and T. D. White, *Science* 203:321 (1979). Also D. C. Johanson and E. Maitley, *Lucy: The Beginnings of Mankind* (New York: Simon & Schuster, 1980).
10. J. E. Cronin et al., *Nature* 292:113 (1981).
11. R. Ardrey, *The Territorial Imperative* (New York: Dell, 1966).
12. R. C. Lewontin, *Annual Review of Genetics* 9:387 (1975).
13. *Scientific Creationism*, p. 178. Full paragraph: "Evolutionists apply evolutionary theory not only to man's origin but also to his later history, interpreting his societies and cultures, and even his economic and political systems, in terms of naturalistic development from one form into another. [Quoted material . . .] since these impinge most directly on man's personal commitments and daily activities."
14. *Ibid.*, pp. 187 ff.
15. *Ibid.*, p. 185.

CHAPTER SIX

1. B. H. Yoo, *Genetical Research* 35:1 (1981).
2. P. Clarke, in M. J. Carlile and J. J. Skehel, ed., *Evolution in the Microbial World* (Cambridge: Cambridge University Press, 1974).
3. R. D. O'Brien, *Insecticides: Action and Metabolism* (New York: Academic Press, 1967).
4. J. H. Hatchett and R. Gallun, *Ann. Ent. Soc. Amer.* 63:1400 (1970).
5. H. C. Bumpus, *Biol. Lec. Mar. Biol. Woods Hole* 11:209 (1899); R. F. Johnston et al., *Evolution* 26:20 (1972).
6. P. T. Boag and P. R. Grant, *Science* 214:82 (1981).
7. H. M. Morris, *Creation: Acts/Facts/Impacts*, p. 45. "Further, such a point-of-view, regardless of the label, is really a contradiction in thought, as well as in terms. Theistic evolution is about as logical as 'Christian atheism' or 'flaming snowflakes.' [Quoted material.]" Edited material reads as in the quote that follows.
8. H. M. Morris, *Ibid.*
9. R. C. Lewontin, *Amer. Natur.* 96:65 (1962).
10. L. Johnson, *Evolution* 36:251 (1982).
11. M. Rose and B. Charlesworth, *Genetics* 97:173 (1981).
12. R. Dawkins, *The Selfish Gene* (New York: Oxford University Press, 1976).
13. W. D. Hamilton, *J. Theoret. Biol.* 31:295 (1971).
14. H. B. D. Kettlewell, *Heredity* 9:323 (1955).
15. R. D. O'Brien, *op. cit.*
16. W. R. Dawson et al., *Evolution* 31:891 (1977).
17. C. Mitter et al., *Evolution* 33:777 (1979).

18. *Scientific Creationism*, p. 70. "In the organic realm, there are many similarities between different kinds of plants and animals, and evolutionists have interpreted these as evidence of common ancestry. Creationists, on the other hand, interpret the same similarities as evidence of common creative planning and design. [Quoted material.]"

19. S. J. Gould, *Ontogeny and Phylogeny*, *op. cit.*

20. W. Durant, *The Story of Philosophy* (New York: Simon & Schuster, 1933).

21. The book *Scientific Creationism* postulates that the original creation was perfect, but that it has been "running down" ever since because of a "basic principle of disintegration" at work in nature. The book says (p. 12) that this is so because any change in a perfect creation must be in the direction of imperfection. Although the authors propose that the Creator designed processes of conservation to maintain His creation, they do not explain why a basic principle of disintegration should also have been instituted.

Beginning with St. Augustine, theologians maintained that disharmony in nature came about because of Adam and Eve's fall from grace. The Venerable Bede declared that before man's fall, animals were harmless: "Fierce and poisonous animals were created for terrifying man (because God foresaw that he would sin), in order that he might be made aware of the final punishment of hell." As late as the eighteenth century, John Wesley declared that before Adam's sin "the spider was as harmless as the fly, and did not lie in wait for blood." Some theologians maintained that volcanic eruptions, earthquakes, and violent storms were manifestations of divine retribution for sin; others declared that they could be caused by devils. St. Thomas Aquinas affirmed that it is "a dogma of faith that the demons can produce wind, storms, and rain of fire from heaven." By the fourteenth century the demonic interpretation led to the belief that storms were the work of witches, who had allied themselves with the devil. In consequence, tens of thousands of people, especially women, were tortured and many thousands killed, in obedience to the scriptural command "Thou shalt not suffer a witch to live." Thus does a literal interpretation of the Bible provide ample explanation of imperfections in the creation.

Clearly, if "scientific" creationism is to provide any explanation of imperfections in organisms or in other features of the universe, it must either attribute some undesirable attributes to the Creator, or invoke subsidiary explanations such as retribution for sin or the operation of devils. Perhaps "scientific creationists" have not explicitly invoked these explanations because of their clear theological content. In any case, such hypotheses clearly cannot pretend to be scientific, and I have not seen any reason to devote space to them.

CHAPTER SEVEN

1. *Scientific Creationism*, p. 15. See note 38, chapter 1.

2. *Ibid.*, p. 62.

3. G. A. Clayton and A. Robertson, *Amer. Natur.* 89:151 (1955).
4. Th. Dobzhansky, *Genetics of the Evolutionary Process* (New York: Columbia University Press, 1970).
5. J. Lederberg and E. M. Lederberg, *J. Bacteriol.* 63:399 (1952).
6. P. Clarke, *op. cit.*
7. F. J. Ayala, *Science* 162:1453 (1968).
8. P. E. Hansche, *Genetics* 79:661 (1975).
9. *Scientific Creationism* p. 56. "As a matter of fact, the phenomenon of a truly beneficial mutation, one which is *known* to be a mutation and not merely a latent characteristic already present in the genetic material but lacking previous opportunity for expression, and one which is permanently beneficial in the natural environment, has yet to be documented. Some evolutionists doubt that they occur at all. [Quotation follows from C. P. Martin, *American Scientist* (January 1953), p. 102.]"
10. *Evolution: The Fossils Say No!* p. 44. Paragraph continues, "Evolutionists claim, however, that a very small fraction (perhaps 1 in 10,000) of these mutations are beneficial. This claim is made, not because we can actually observe such favorable mutations occurring, but because evolutionists know that unless favorable mutations do occur, evolution is impossible. In the final analysis, all of evolution must be ascribed to mutations."
11. J. Antonovics et al., *Adv. Ecol. Res.* 7:1 (1971).
12. Th. Dobzhansky, F. J. Ayala, G. L. Stebbins, and J. W. Valentine, *Evolution* (San Francisco: W. H. Freeman, 1977).
13. G. A. Clayton and A. Robertson, *op. cit.*
14. *Scientific Creationism*, pp. 56–57. "That the net effect . . . removed from the environment" is followed by a quote from *Scientific American*: "The most important actions that need to be taken, however, are in the area of minimizing the addition of new mutagens to those already present in the environment. Any increase in the mutational load is harmful, if not immediately, then certainly to future generations." *Scientific Creationism* continues, "It does seem that, if evolutionists really believed that evolution is due to mutations, they would favor all measures which could increase the rate of mutations and thus facilitate further evolution. Instead, they have consistently for decades opposed nuclear testing for the very purpose of *preventing* mutations!"
15. R. F. Johnston and R. K. Selander, *Science* 144:548 (1964).
16. P. A. Phillips and M. M. Barnes, *Ann. Ent. Soc. Amer.* 68:1053 (1975).
17. *Evolution: The Fossils Say No!*, p. 45.
18. R. K. Selander, *Amer. Zool.* 10:53 (1970).

CHAPTER EIGHT

1. A. O. Lovejoy, *op. cit.*
2. In D. L. Hull, *op. cit.*
3. J. C. Greene, *op. cit.*

4. *Evolution: The Fossils Say No!*, p. 37. "Of greatest importance to our discussion, however, is the fact that no significant evolutionary change has occurred in these moths. These moths today not only are still moths, *but they are still peppered moths, Biston betularia.* This evidence, therefore, is irrelevant to the questions we seek to answer: did these lepidopterous insects arise by a naturalistic, mechanistic process from a nonlepidopterous insect? Did the insects themselves arise from a noninsect form of life?"

5. *Ibid.*, p. 38.

6. *Ibid*, p. 39. "In summary, then, by evolution we mean a process which is supposed to have been responsible for converting the most primitive form of life, the hypothetical primordial cell, via innumerable increasingly complex forms of life, into man, the highest form of life. The theory of evolution, then, proposes that basically different types of plants and animals have arisen from common ancestors, which in turn had arisen from more ancient and more primitive forms of life. [Quoted material.]"

7. *Ibid.*, p. 32. "We must here attempt to define what we mean by a basic kind. [Quoted material.] All humans, for example, are within a single basic kind, *Homo sapiens.* In this case, the basic kind is a single species."

8. *Ibid.*, p. 35. See Chapter 1, note 36.

9. C. Darwin, *The Origin of Species.*

10. See, e.g., E. Mayr, *Animal Species and Evolution* (Cambridge, Mass.: Harvard University Press, 1963).

11. R. B. Goldschmidt, *op. cit.*

12. L. Ehrman, *Evolution* 14:137 (1965).

13. Th. Dobzhansky and O. Pavlovsky, *Nature* 23:289 (1971).

14. E. Zimmerman, *Evolution* 14:137 (1960).

15. G. Fryer and T. D. Iles, *The Cichlid Fishes of the Great Lakes of Africa* (Neptune City, N.J.: T.F.H. Publications, 1972).

16. H. L. Carson et al., in M. K. Hecht and W. C. Steere, eds., *Essays in Evolution and Genetics in Honor of Theodosius Dobzhansky* (New York: Appleton-Century-Crofts, 1970).

17. N. Eldredge and S. J. Gould, *op. cit.*

18. W. J. Bock, *Evolution* 24:704 (1970).

19. See, for example, S. M. Stanley, *The New Evolutionary Timetable* (New York: Basic Books, 1981).

20. G. L. Stebbins, *Flowering Plants: Evolution Above the Species Level* (Cambridge, Mass.: Harvard University Press, 1974).

21. J. Clausen, D. D. Keck, and W. M. Hiesey, *Amer. Natur.* 81:114 (1947).

22. See, for example, M. Pei, *The Story of Language* (Philadelphia: Lippincott, 1949).

CHAPTER NINE

1. Z. A. Medvedev, *The Rise and Fall of T. D. Lysenko* (New York: Columbia University Press, 1969).

2. *Scientific Creationism*, p. 15. Quoted material is part of a list of objections to teaching evolution exclusively, which begins as quoted in note 38 to Chapter 1, and continues, "5. It is believed by creationists to be harmful to the child or teenager since it contradicts his innate consciousness of reality and thus tends to create mental and emotional conflicts within him. 6. It tends to remove all moral and ethical restraints from the student and leads to an animalistic amoralism in practice. 7. It may tend to rob life of meaning and purpose in view of the implanted concept that the student is merely a chance product of a meaningless, random process. 8. Evolutionary philosophy often leads to a conviction that might makes right, leading either to anarchism (uncontrolled evolution) or collectivism (controlled evolution)."

3. K. R. Popper, *Conjectures and Refutations* (New York: Harper & Row, 1963), p. 6.

4. *Ibid.*; see also K. R. Popper, *The Logic of Scientific Discovery* (New York: Harper & Row, 1968). Sir Karl Popper is widely considered one of the outstanding and influential philosophers of science, and is primarily responsible for articulating the view that science progresses primarily by posing hypotheses that can be shown wrong if in fact they are false.

5. S. J. Gould, *The Mismeasure of Man* (New York: Norton, 1981).

6. R. Lewin, *Science* 213:316 (1981).

7. D. D. Dorfman, *Science* 201:1177 (1978).

8. In D. L. Hull, *op. cit.*, p. 9.

9. P. B. Medawar, *The Art of the Soluble* (London: Methuen, 1967).

10. See note 4.

11. This is not a purely personal statement; see, for example, R. C. Lewontin, *BioScience* 31:559 (1981) for a similar declaration by an eminent geneticist.

12. R. C. Lewontin, *Nature* 236:181 (1972).

13. K. R. Popper, *Dialectica* 32:339 (1978). See also K. R. Popper, *New Scientist* 87:611 (1980).

14. W. W. Benson, *Science* 176:936 (1972).

15. S. J. Gould, *Science* 216:380 (1982); also *Paleobiology* 6:96 (1980).

16. D. T. Gish, *Impact*, no. 43 (San Diego: Institute for Creation Research, 1977).

17. *Evolution: The Fossils Say No!*, p. 11. "The creation model, on the other hand, postulates that all basic animal and plant types (the created kinds) were brought into existence by acts of a supernatural Creator using special processes which are not operative today."

18. *Ibid.*, p. 40. The elided material is "This is why we refer to creation as special creation." However, in more general usage, "special creation" means separate creation of each species, not the meaning Gish attributes to the term.

CHAPTER TEN

1. L. W. Alvarez, W. Alvarez, F. Asaro, and H. V. Michel, *Science* 208: 1095–1108 (1980); see also R. Ganapathy, *Science* 209:921–23 (1980), and K. J. Hsü et al., *Science* 216:249–56 (1982).

2. See for example, R. A. Kerr, *Science* 210:514–17 (1980).

3. D. T. Gish et al., *Impact*, no. 95 (San Diego: Institute for Creation Research, 1981).

4. *Evolution: The Fossils Say No!*, p. 21.

5. *Scientific Creationism*, p. 19; See Chapter 1, note 21.

6. E. Mayr and W. B. Provine, *op. cit.*

7. *Scientific Creationism*, pp. 121–22.

8. H. M. Morris, *Impact*, no. 77 (San Diego: Institute for Creation Research, 1979).

9. R. C. Lewontin, "Adaptation," *Scientific American* 239 (3):212–30 (1978).

10. H. M. Morris, *Impact*, no. 77.

11. *Scientific Creationism*, p. 95.

12. D. M. Raup, *Science* 213:289 (1981).

13. T. H. Jukes, *Trends in Biochemical Sciences* 6 (7):1–2 (1981).

14. Clark Summit, Pennsylvania, August 5, 1981.

15. H. M. Morris, *Introducing Scientific Creationism into the Public Schools* (San Diego: Institute for Creation Research, 1975); H. M. Morris et al., *Creation: Acts, Facts, Impacts* (San Diego: Institute for Creation Research, 1974), pp. 157 ff.

16. *Creation: Acts, Facts, Impacts*, p. 160. Quoted material is preceded by "In a day and age which practically worshipped at the shrine of scientific progress, as was true especially during the century from 1860 to 1960, such universal scientific racism was bound to have repercussions in the political and social realms. The seeds of evolutionary racism came to fullest fruition in the form of National Socialism in Germany."

17. P. Cloud, " 'Scientific Creationism'—a New Inquisition," *The Humanist* (January–February 1977).

18. *Scientific Creationism*, p. 92. "Nevertheless, it is true that the evolution model is fundamentally tied to uniformitarianism, since it assumes that present natural laws and processes suffice to explain the origin and development of all things. The creation model [quoted material]. It centers its explanation of past history around both a period of special *constructive* processes and a period of special *destructive* processes, both of which operated in ways or at rates which are not commensurate with present processes."

19. *Ibid.*, p. 133. Paragraph continues, "Also, as we shall see in the next section, there are so many sources of possible error or misinterpretation in radiometric dating that most such dates are discarded and never used at all, notably whenever they disagree with the previously agreed-on dates."

20. *Ibid.*, p. 137. Paragraph begins, "In attempting to determine the real age of the earth, it should always be remembered, of course, that recorded history began only several thousand years ago."

21. *Ibid.*, p. 40.

22. Isaac Asimov, "The 'Threat' of Scientific Creationism," *New York Times Magazine*, June 14, 1981.

23. D. T. Gish, lecture, Clark Summit, Pennsylvania, August 5, 1981. See *Scientific Creationism*, pp. 59–69.

24. D. T. Gish, *Impact*, no. 31 (San Diego: Institute for Creation Research, 1976).

25. *Evolution: The Fossils Say No!*, p. 44. Paragraph continues with, "It is doubtful that, of all the mutations that have been seen to occur, a single one can definitely be said to have increased the viability of the affected plant or animal," and then with the material quoted in note 10, chapter 7.

26. *Scientific Creationism*, pp. 52, 53. Paragraph continues, "Nevertheless, this phenomenon of recombination followed by natural selection is somehow regarded by evolutionists as a very important part of their model."

27. H. B. D. Kettlewell, *Annual Review of Entomology* 6:245 (1961).

28. E. B. Ford, *Ecological Genetics* (London: Chapman and Hall, 1971).

29. *Scientific Creationism*, p. 51. "Normal variations were later found to be subject to the rigid laws of Mendelian inheritance, representing nothing really novel but only characters already latent within the genetic system. [Quoted material.] Variation is horizontal, not vertical!"

30. *Evolution: The Fossils Say No!*, p. 49. Paragraph begins, "We find fossils of crossopterygian fishes which are alleged to have given rise to the amphibia. We find fossils of the so-called 'primitive' amphibia. Since the transition from fish to amphibia would have required many millions of years, during which many hundreds of millions, even billions, of the transitional forms must have lived and died, many of these transitional forms should have been discovered in the fossil record even though only a minute fraction of these animals have been recovered as fossils."

31. *Ibid.*, p. 62. Elided material is "In these sedimentary deposits are found billions and billions of fossils of highly complex forms of life. These include sponges, corals, jellyfish, worms, mollusks, crustaceans; in fact, every one of the major invertebrate forms of life have been found in Cambrian rocks. [Quoted material.] Certainly it can be said without fear of contradiction that the evolutionary ancestors of the Cambrian fauna, if they ever existed, have never been found."

32. P. Cloud, *op. cit.*

33. D. T. Gish, *Evolution: The Fossils Say No!*, pp. 84, 85. "Thus, in not a single instance concerning origin of flight can a transitional series be documented, and in only one case has a single intermediate form been alleged. In the latter case, the so-called intermediate is no real intermediate at all because,

as paleontologists acknowledge, *Archaeopteryx* was a true bird—it had wings, it was completely feathered, it *flew* (see Fig. 3). It was not a half-way bird, it *was* a bird." "While modern birds do not possess teeth . . . [quoted material] some reptiles have teeth while some do not. Some amphibians have teeth, but some do not. In fact, this is true throughout the entire range of the vertebrate subphylum—fishes, Amphibia, Reptilia, Aves, and Mammalia, inclusive."

34. *Ibid.*, p. 80. "The two most easily distinguishable osteological differences between reptiles and mammals, however, have never been bridged by transitional series. All mammals, living or fossil, have a single bone, the dentary, on each side of the lower jaw, and all mammals, living or fossil, have three auditory ossicles or ear bones, the malleus, incus, and stapes. In some fossil reptiles the number and size of the bones of the lower jaw are reduced compared to living reptiles. Every reptile, living or fossil, however, has at least four bones in the lower jaw and only one auditory ossicle, the stapes. There are no transitional forms showing, for instance, three or two jaw bones, or two ear bones. No one has explained yet, for that matter, how the transitional form would have managed to chew while his jaw was being unhinged and rearticulated, or how he would hear while dragging two of his jaw bones up into his ear."

35. This discussion is based on, and the quotations are taken from, E. H. Colbert, *Evolution of the Vertebrates* (New York: Wiley, 1955). See pp. 121 ff.

36. *Evolution: The Fossils Say No!*, p. 93.

37. G. L. Stebbins and F. J. Ayala, *Science* 213:967 (1981).

38. *Scientific Creationism*, p. 53. Paragraph continues, "Yet, somehow, if the evolution model is valid, wings have 'evolved' four different times (in insects, flying reptiles, birds and bats) and eyes have 'evolved' independently at least three times. Salisbury has recently commented on this remarkable fact as follows: [quotation here from *American Biology Teacher* (September 1971), p. 338]."

39. See, e.g., R. M. Eakin, "Evolution of Photoreceptors," in Th. Dobzhansky, M. K. Hecht, and W. C. Steere, eds., *Evolutionary Biology*, vol. 2 (New York: Appleton-Century-Crofts, 1968); B. Rensch, *Evolution Above the Species Level* (New York: Columbia University Press, 1959).

40. H. M. Morris, *Impact*, no. 77; D. T. Gish, *Impact*, nos. 42 and 43 (San Diego: Institute for Creation Research, 1976 and 1977).

41. D. T. Gish, *Impact*, no. 43.

42. S. M. Stanley, *Proc. Nat. Acad. Sci. U.S.A.* 72:646 (1975).

43. D. T. Gish, *Impact*, no. 43.

44. The basis for testing the hypothesis is explained by S. J. Gould, in R. D. Milkman, ed., *Perspectives on Evolution* (Sunderland, Mass.: Sinauer, 1982).

45. *Evolution: The Fossils Say No!*, p. 23.

46. A. D. White, *op. cit.*, pp. 114–42.

CHAPTER ELEVEN

1. *Scientific Creationism*, p. 20. "We conclude from the law of cause-and-effect that the First Cause of all things must be an infinite, eternal, omnipotent, omnipresent, omniscient, moral, spiritual, volitional, truthful, loving, living Being! Do such adjectives describe Matter? Can random motion of primeval particles produce intelligent thought or inert molecules generate spiritual worship? To say that Matter and its innate properties constitute the ultimate explanation for the universe and its inhabitants is equivalent to saying that the Law of Cause-and-Effect is valid only under present circumstances, not in the past."
2. *Ibid.*, p. 12. See Chapter 1, note 20.
3. Th. Dobzhansky, *Genetics and The Origin of Species* (New York: Columbia University Press, 1951).
4. *Scientific Creationism*, pp. 32, 35. See Chapter 1, notes 23 and 24.
5. *Ibid.*, p. 118. Paragraph begins, "The above of course is only the barest outline of the great variety of phenomena that would accompany such a cataclysm."
6. *Ibid.*, pp. 118 ff.

CHAPTER TWELVE

1. The history of Social Darwinism and the quotations given here can be found in R. Hofstadter, *Social Darwinism in American Thought* (Boston: Beacon Press, 1955).
2. See, for example, M. Harris, *Culture, People, Nature: An Introduction to General Anthropology* (New York: Thomas Y. Crowell, 1975).
3. R. C. Lewontin, *Annual Review of Genetics* 9:387 (1975).
4. S. Toulmin, in S. Toulmin, R. W. Hepburn, and A. MacIntyre, eds., *Metaphysical Beliefs* (London: SCM Press, 1957); J. Collins, *Thought* 34:185 (1959).
5. J. Huxley, *The Romanes Lecture 1943*, in T. H. Huxley and J. Huxley, *Evolution and Ethics 1893–1943* (London: Pilot Press, 1947).
6. T. H. Huxley, *The Romanes Lecture 1893*, in T. H. Huxley and J. Huxley, *op. cit.*
7. *Evolution: The Fossils Say No!*, p. 40. See note 18 to Chapter 9 for rest of paragraph.

INDEX

This index is not intended to be complete. Only key concepts, historical figures, and biological terms are included. Reference is given to first entries and, subsequently, only to the most prominent places in which the concepts are discussed.

ABOUT THE AUTHOR

Douglas Futuyma received his B.S. from Cornell University and his M.A. and Ph.D. from the University of Michigan at Ann Arbor. He is the author of numerous articles in the leading biological and ecological journals in his field, the author of a widely used college textbook on evolution, *Evolutionary Biology*, and editor of *Evolution*, the international journal of the Society for the Study of Evolution. At present, he is an associate professor in the Department of Ecology and Evolution at the State University of New York at Stony Brook.